Synthesis Lectures on Computation and Analytics

This series focuses on advancing education and research at the interface of qualitative analysis and quantitative sciences. Current challenges and new opportunities are explored with an emphasis on the integration and application of mathematics and engineering to create computational models for understanding and solving real-world complex problems. Applied mathematical, statistical, and computational techniques are utilized to understand the actions and interactions of computational and analytical sciences. Various perspectives on research problems in data science, engineering, information science, operations research, and computational science, engineering, and mathematics are presented. The techniques and perspectives are designed for all those who need to improve or expand their use of analytics across a variety of disciplines and applications.

Jay Wang · Adam Wang

Introduction to Computation in Physical Sciences

Interactive Computing and Visualization with Python™

 Springer

Jay Wang
Dartmouth, MA, USA

Adam Wang
Nashville, TN, USA

ISSN 2766-8975   ISSN 2766-8967   (electronic)
Synthesis Lectures on Computation and Analytics
ISBN 978-3-031-17648-7   ISBN 978-3-031-17646-3   (eBook)
https://doi.org/10.1007/978-3-031-17646-3

This Springer imprint is published by the registered company Springer Nature Switzerland AG
The registered company address is: Gewerbestrasse 11, 6330 Cham, Switzerland

Preface

This book introduces computation and modeling of problems in physical sciences with emphasis on interactive just-in-time computing, data analysis, and effective visualization. Computational modeling has long been an established practice in science and engineering and is regarded as the third pillar alongside experimentation and theory. It is also becoming increasingly inseparable and integral in a wide range of industry sectors. In education, computational thinking—broadly construed as the mindset and skillset of problem-solving with computational methodologies including modeling, abstraction, symbolic and algorithmic representations, and numeric processing—is being embraced as a standard component, and even a requirement, at all levels.[1]

Experience and evidence from physics educators and physics education research have shown that computation and integration thereof into physics has many positive benefits including increasing student engagement, active learning, and improved learning outcomes and understanding. However, there are also well-documented challenges to effective implementation in terms of pedagogical consideration and practical constraints. We developed this book with the goal of lowering such barriers. In terms of coverage, we have carefully selected a set of topics, drawn from experience and best practice in the physics education community, that is pedagogically significant and also reflective of the current trend in academia and in industry. In terms of access to computing resources, we have adopted a COD (computing on demand) and JIT-C (just-in-time computing) model where nearly all programs are cloud-based, requiring no software installation of any kind for easily accessing or running them in a web browser on any machine, especially on mobile devices. Running a simulation, or sharing one, is just a click away. We designed most programs as minimally working programs (MWP) as part of intentional scaffolding to help beginners develop computational problem-solving and thinking skills. MWPs are effective pedagogically to teaching and learning computational modeling. They may be used as basis templates to be modified, extended, or combined for further exploration in projects and exercises.

---

[1] See J. M. Wing, Computational thinking, *Communications of the ACM*, **49**, 33–35 (2006).

The book is broadly organized into two parts. Part one consists of Chaps. 1 to 4 and describes basic programming and tools, including programming environments, Python tutorial, interactive computing, and elementary algorithms. Part two includes Chaps. 5 to 9 and focuses on application of computational modeling of select problems in fundamental areas of physical and data science including classical mechanics, modern physics, thermal and complex systems, as well as from contemporary topics including quantum computing, neural networks and machine learning, and global warming and energy balance.

The materials are arranged as follows. In Chap. 1, we introduce programming and development environments including Jupyter notebook—the recommended environment, its cloud-based counterpart Binder, a close variant Colab, the native Python IDLE—integrated development and learning environment, Spyder, and sharing-friendly GlowScript and Trinket. We also describe the online program repository, how to access and run programs in the cloud, and optional software installation to run programs locally. An interactive Python tutorial is given in Chap. 2 on basic programming elements, including the important aspect of error handling, debugging, and troubleshooting. Next, Chap. 3 focuses on interactive computing and visualization methods with Python widgets and VPython animation libraries. Key scientific computing libraries are also described with interactive examples including symbolic and numeric computation with Sympy, Numpy, Scipy, as well as Numba for optimization. Rounding out part one, we introduce in Chap. 4 common elementary algorithms such as finite difference, numerical integration including that of differential equations, finding of roots and minima or maxima, and curve fitting.

We begin part two with Chap. 5 on force and motion where we pedagogically introduce kinematic concepts, and discuss projectile motion with and without air resistance, harmonic oscillation, planetary motion, and motion in electromagnetic fields. In Chap. 6 we discuss classical coupled oscillations and wave propagation and introduce linear systems and fundamental modes, leading up to simulations of matter waves in modern and quantum physics next in Chap. 7 where we consider relativity of space and time, diffraction and double-slit interference, formation of discrete quantum states and its visualization. We also introduce quantum information processing and quantum computing with an interactive quantum simulator and a step-by-step walkthrough of a nontrivial real-world example: quantum search—the Grover algorithm. We consider random processes and thermal physics in Chap. 8 such as nuclear decay, Brownian motion, Monte Carlo simulations of thermal energy distribution and spin systems. We close the chapter with a discussion of greenhouse effect, energy balance, and global warming. Lastly, we discuss complex systems in Chap. 9 including cellular automata, traffic modeling, and nonlinear dynamics and chaos. We conclude with an introduction to neutral network and machine learning, and describe a worked example with details in the training and prediction of game of life.

In terms of background and preparation, readers without prior programming experience should start with part one, focusing on the interactive tutorial (Chap. 2) and moving on to Python libraries (Chap. 3) and basic algorithms (Chap. 4). To get the most out of part two, the reader should be calculus-ready, having studied precalculus if not calculus already. We took care to make the book as self-contained as possible in terms of explaining physical concepts necessary for a particular simulation. Even so, a basic understanding of physics at introductory level is very helpful such as familiarity with basic concepts of motion and Newton's second law. We recommend covering Chap. 5, not necessarily in its entirety, before Chap. 6. The other individual chapters are sufficiently decoupled from each other that they may be used as separate modules and studied independently.

Integral to the main content, the book contains extensive sets of carefully designed and tested exercises at the end of each chapter. Most exercises are closely related to the MWPs discussed in the main text, and can be used for further exploration of a given problem, as project assignments or self-paced study. Exercises vary in depth and difficulty levels, and the more challenging ones are marked by asterisks.

We thank Susanne Filler, Executive Editor at Springer Nature, and her team for the assistance and timely guidance in this project. JW is grateful to his colleagues at UMass Dartmouth and in the PICUP community (Partnership for Integration of Computation into Undergraduate Physics, http://gopicup.org) for their support and inspiration; and to the countless students who participated in research projects or through coursework and provided feedback in various ways: you have had more impact on this project than you ever knew. Finally, we are fortunate to have the unwavering encouragement, understanding, and patience of our loved ones: Huihua, Kathy, and Greg. Thank you.

Dartmouth, USA   Jay Wang
Nashville, USA   Adam Wang

Contents

Programming Environments

1

Computing hardware has changed dramatically from the vacuum-tube computers of 1950s, to shared mainframes of 1980s, and now to portable devices of present with computing power that far exceed the supercomputers of 1990s. Correspondingly, computational environments have also evolved from central computing in the early days to distributed computing, and from text-based terminal interfaces to graphical and web- or cloud-based development environments at present. We briefly describe several programming environments and take a side by side glance over Jupyter notebook, Colab, Spyder, and so on. Each has its pros and cons, and the choice often comes down to personal preference, familiarity, or workflow. We recommend the Jupyter notebook environment because it is an all-around good environment: friendly interface, easy to use, selective code-cell execution, self-documenting with code, graphics, LATEX style text, and readily suitable for project reports. See Table 1.1 for a quick summary of program types and suitable environments.

1.1 Jupyter Notebook Environments

We first describe browser-based programming environments for developing and running Jupyter notebook programs that end with the file extension `.ipynb`.

1.1.1 Jupyter Notebook

Jupyter notebook (or simply Jupyter) is a browser-based graphical programming development environment (the latest version is called JupyterLab). It is included in all-in-one

© The Author(s), under exclusive license to Springer Nature Switzerland AG 2023
J. Wang and A. Wang, *Introduction to Computation in Physical Sciences*, Synthesis Lectures on Computation and Analytics,

Table 1.1 Program type and development environment

Program type	Environment	URL
Notebook (.ipynb)	Jupyter/JupyterHub	https://jupyter.org/try, https://jupyter.org/hub
	Binder	https://github.com/com-py/intro, https://mybinder.org
	Colab	https://colab.research.google.com
Standard (.py)	IDLE, Spyder, terminal	In terminal mode, edit (e.g. with notepad++) and run programs with command python prog.py https://notepad-plus-plus.org/
GlowScript	GlowScript, Trinket	https://www.glowscript.org, https://trinket.io

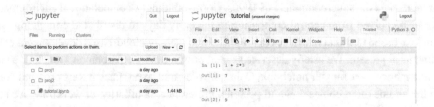

Fig. 1.1 Jupyter notebook environment. Left: home screen; right: program screen

packages such as Anaconda[1] or can be custom-installed. One can also try it out on the web (see Table 1.1). It is also possible to run Jupyter notebook programs remotely using Jupyter-Hub (see end of this section). Jupyter notebook supports multiple programming languages, although we will focus on Python.

Figure 1.1 shows the basic layout of Jupyter notebook. The home screen has two tabs listing files and folders (from the specified home directory of Jupyter when it was launched) and programs currently running, respectively.[2] Clicking on a folder opens that folder, and clicking a program launches the program in a separate tab. Selecting an item initiates a set of action buttons to rename, copy, or delete the item. Above the folders and files to the right are two buttons labeled Upload and New. We can upload a program (or file) on the local machine with the Upload button. This is useful if you want to run a program found elsewhere, one from this book for example. You download it and the upload to Jupyter to

---

[1] https://www.anaconda.com.
[2] The cluster tab would show parallel processing information that is not applicable to our discussion.

run it. The New buttons opens a menu where we can create a new program or folder, or open a terminal window. We can shut down Jupyter and quit with the Quit button.

To launch a program from the home screen, click on tutorial. The browser will switch to a new tab, the program screen (Fig. 1.1, right). There are a number of standard menu items and icons below them to perform various tasks (mouse over each icon to see a popup description). The help menu can be quite handy, and brings us to external links to learn more about a topic.

The main work area beneath show either code or markdown cells. A currently selected cell can be executed by clicking on the run icon or by shift-enter (holding down shift while pressing the enter key). The output (if any) is shown below, and the next cell is selected (highlighted). Graphics output will be automatically saved.

A cell type can be changed from the dropdown box (only code or markdown will be of interest to us). The markdown cell type allows for any text or documentation including LaTeX formatted mathematical expressions, embedded graphics or links to websites. Learn more about it by selecting markdown from the help menu.

We should keep in mind that Jupyter notebook remembers all the variables and their current values at the point of execution. Sometimes this memory has unintended consequences (left over or improperly initiated variables). To clear Jupyter of memory effects, choose from the Kernel menu and restart the kernel. If a program appears unresponsive as sometimes happens with VPython graphics, or program output seems wrong, try to fix it by restarting the kernel.

JupyterHub

If one wishes to run Jupyter notebook programs without software installation by the end user, there are two possibilities. The first option is to install JupyterHub (see Table 1.1) on your own server or other hosting service. A user can access Jupyter notebook remotely and run programs in a web browser. The JupyterHub setup is a popular choice at many institutions. The second option is to use a cloud service like Colab (see Sect. 1.1.2).

Binder

Most of Jupyter notebook programs in this book can be run from Binder, a platform that converts Github repositories into interactive notebooks on demand. There is no need to install anything. Simply go to the book repository on Github[3] (also see Table 1.1), click on a program, and follow the instructions and the link.

---

[3] https://github.com/com-py/intro.

Fig. 1.2 Colab programming
environment

1.1.2   Colaboratory

Colaboratory (or just Colab) is a repackaged Jupyter platform offered by Google (see Table 1.1) that runs Python notebook programs in a web browser.[4] The basic service is free, and requires no installation of software.

Colab imitates the Jupyter user interface (Fig. 1.2) but with significant alterations in layout and functionality. The advantages are that access is simple, requiring only a Google account without any software installation or configuration. One can open a notebook program with a shared link or click on one stored on Google drive. The latter will launch Colab in the browser. Members of a group can collaborate and leave comments.

The disadvantages are that Colab can be sluggish because resources are not guaranteed with the free service. The altered user interface and functionality could be a hindrance to those familiar with the standard Jupyter operations. Colab has not and does not support VPython programs. Also, data privacy and where it is hosted may be a concern to some users.

The biggest drawback for some users may be the limitations (and whims) of Colab beyond users' control. One of the limitations is the lack of support of certain libraries. For example, VPython, a popular 3D graphics and animation library used in education and research, does not work in Colab, though it works fine in Jupyter. If one wishes to run Jupyter notebook programs in the cloud, consider the JupyterHub or the Binder option (see Sect. 1.1.1).

1.2   Integrated Development Environments

We describe two integrated development environments for developing and running standard Python programs with the extension .py. Both come with a program editor and an integrated debugger.

---

[4] Microsoft had a similar service called Azure Notebooks now renamed Azure ML Studio. See https://studio.azureml.net.

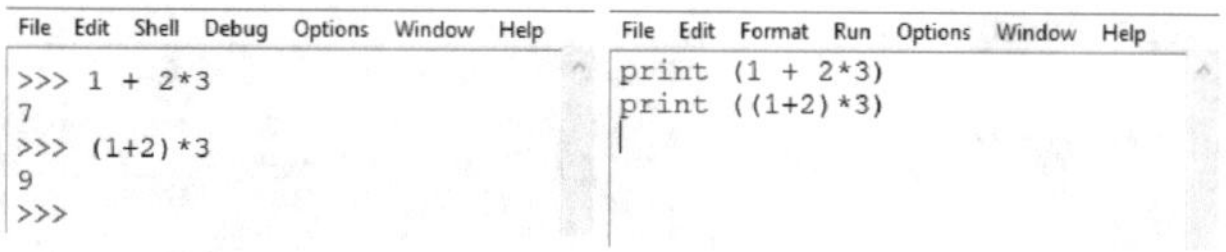

Fig. 1.3 IDLE programming environment. Left: shell window; right: editor window

1.2.1   IDLE

The IDLE (integrated development and learning environment) is a basic programming environment provided and automatically installed with Python or all-in-one packages like Anaconda. It includes an interactive shell, a program editor, and a debugger (Fig. 1.3).

When the IDLE is launched, the shell window starts (Fig. 1.3, left). We can type a Python statement at the prompt ≫, and execute it with the enter key. This would be an ideal place to learn or use Python as a calculator (see Sect. 2.1). Text output is written after the prompt within the shell. Graphics output such as Matplotlib is displayed in a separate popup window or in a browser tab for VPython output (see Sect. 3.2.2).

One can open an existing program (.py type) or start a new one with the program editor (Fig. 1.3, right). When ready (after saving changes), we can run the program from the menu. Text output will appear in the shell window and graphics output in external windows as usual.

To use the debugger, turn it on from the shell window menu. A separate debugger window will appear. Now if a program is run from the editor, the debugger allows for stepping through the program line by line, setting break points, or examining variables (Spyder has a more advanced debugger, see Sect. 1.2.2 next). For most short programs we deal with here, a more practical way of error fixing is to place the print function at key locations to look at the contents of a variable, and use it as clues to deduce potential problems (more on error fixing in Sect. 2.11.4).

1.2.2   Spyder

Like the IDLE, Spyder is an integrated development environment for Python. But its interface is highly customizable. At its core, it includes a program editor, launcher, and a debugger. Figure 1.4 shows a particular layout with the program editor, an output window, and a file window. It is prepackaged with Anaconda and may also be installed separately.

Standard menu items are at the top of the interface. Immediately below are several sets of icons to perform tasks including opening or saving files in the editor, launching programs, debugging, and so on.

As with the IDLE, Spyder can run only standard Python programs (.py, not Jupyter notebook .ipynb), and it does not output the result like a calculator the way Jupyter

Fig. 1.4 Spyder programming
environment. Left: text editor;
middle: output window; right:
file window

does. Instead, we must use the `print` function for text output. However, Spyder also has
internal hooks to intercept and redirect graphical output such as from Matplotlib to the output
window. VPython animation will appear in a web browser window and not in-page as with
Jupyter notebook (see Chap. 3).

By default, Spyder runs the whole program. But it is possible to selectively execute
statement blocks or cells marked with special directives (#%%). It has an advanced debugger
that enables one to step through statements, set break points where code execution is paused,
and inspect contents of variables. As stated earlier, for our purpose, most errors can be traced
and fixed with the help of `print` function (see error fixing, Sect. 2.11.4).

1.3   GlowScript and Trinket

GlowScript
GlowScript (renamed to Web VPython now) is a programming environment for creating 3D
animations. It is a free service running in the cloud calling the VPython graphics library that
also runs in a standard local Python installation (see Sect. 3.2.2).

Any device, be it a laptop or a smart phone, can run GlowScript programs because they
run in a standard web browser (see Program A.7.1 for an example). GlowScript programs
can be accessed with a shared link, or they can be created in a web browser after setting up
a free account.

The GlowScript interface is simple. It has two modes, the editor mode and the execution
mode (output). One can switch between the two with a single click. Only the owner of
a program can make changes or create sharable links. Nonowners can run the program,
download it, or make changes to a copy.

Trinket
Trinket is another programming environment in the cloud that can run GlowScript pro-
grams unmodified (plus programs in Python and several other programming languages).
Like GlowScript, it runs in a web browser with a simple interface (Fig. 1.5).

Figure 1.5 shows a GlowScript program in Trinket (the same program animating an
oscillating ball discussed later in Sect. 3.2.2). Trinket has a couple of unique features. One

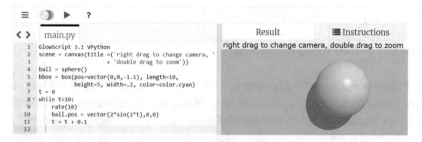

Fig. 1.5 Trinket (GlowScript) programming environment. Left: program editor; right: output window. The output is post-processed with POV-Ray to include ray tracing effects

is that it displays the code and the output side-by-side in the browser. making it easy to compare changes to the code and the difference in the outcome. Another is that Trinket can save changes to a program and create a new sharable link, even if a user is not the owner of the originally shared program and does not have a Trinket account. This makes sharing really easy (of course, it is recommended to create an account to keep track of programs and changes made).

The biggest advantage of GlowScript and Trinket is that both enable nonspecialist users to easily incorporate 3D animation to enhance the simulations that can be shared just as easily. The disadvantage, though, is that we cannot use standard Python libraries in these environments.[5] This limitation is not as severe for sharing short simulations or demonstrations.

Summary

We have briefly introduced several programming environments. Table 1.1 lists the program types and suitable development environments.

Each environment has its advantages and disadvantages relative to the others. As stated earlier, depending on your workflow, choose the ones according to your personal preference, familiarity, and efficiency. More likely, you will find a combination of several environments to be useful.

---

[5] For non-GlowScript, standard Python programs, Trinket supports a limited set of libraries to be imported such as Matplotlib, Numpy and a few more, but not many others like Scipy or Sympy.

1.4   Program Access and Installation

Accessing and running programs

Most Python programs to be discussed in the following chapters are given in both standard
Python format (extension .py) and Jupyter notebook format (extension .ipynb). The two
formats are functionally the same in terms of core computation, namely methods and results,
but are different in terms of appearance and presentation owing to the different environments.
These programs are accessible in several different ways.

- Program listing: Appendix A
  Standard Python format (.py) programs are listed in Appendix A for easy reference and
  study aid. We provide discussions and explanations in the text utilizing Jupyter notebook
  cell structure for code snippets or short but full programs.
- Online repository: https://github.com/com-py/intro
  All the programs discussed in the text are accessible through the online repository. It is
  recommended to use online access because it will be actively updated as warranted. Furthermore, where appropriate, a VPython version of a program is also provided, enabling
  3D animation in the simulation (see Sect. 3.2.2). As a result, some programs have four
  functionally-equivalent versions: standard and notebook formats, each with and without
  VPython enabled.
  A user can run Jupyter notebook programs on a just-in-time basis in-situ at the online
  repository because they are linked to Binder (Table 1.1). It requires just a click, and no
  download or installation of any kind is necessary.
  To run a program locally, follow the instructions at the repository to download and save
  the program, and run it from your machine. For notebook programs, upload it to Jupyter
  to examine and run it (Sect. 1.1.1).

For the purpose of demonstration and easy sharing, some programs are also given in
supplementary GlowScript and Trinket formats. They are available online.

- GlowScript set: https://www.glowscript.org/#/user/jaywang/folder/intro/
  The link contains a set of programs closely related to the counter parts such as Pro-
  gram A.3.4 and Program A.7.1. The set is continually updated. Anyone can view,
  run, share, or download GlowScript programs. The downloaded programs are standard
  VPython programs that can be run locally.
- Trinket set: Online repository
  A set of Trinket programs may be found at the online repository discussed earlier. Like
  GlowScript programs, one can run and share them freely.

Installation

As stated earlier, the programs presented in this book can be accessed and run from the online repository without any software installation. Still, in many cases, a local installation is preferred or required.

The easiest method is to install the all-in-one Python distribution Anaconda mentioned earlier.[6] Download the installer from the URL (Sect. 1.1.1) and run it. The Anaconda installation includes everything we need (and much more) with one notable exception, unfortunately: it does not include VPython by default. To install VPython afterward, open an Anaconda terminal (different than the DOS terminal on Windows), use one of the commands

```
pip install vpython
```

or

```
conda install −c vpython vpython
```

Now, this will complete the installation with all we need.

The next easiest method is to install Python and associated libraries manually. First, install base Python.[7] Second, download the file `requirements.txt` from the online repository given earlier. Then open a system terminal and issue the command

```
pip install −r requirements.txt
```

This will complete a custom Python installation for our every need. Should an error such as `pip is not found` occur, prefix `pip` with the right path where base Python is installed into. On Windows, for example, this could be

```
C:\Python3\Scripts\pip install −r requirements.txt
```

Now we have everything ready, go forth and simulate. If you are new to Python or need a refresher, start with the interactive tutorial in Chap. 2. However, like driving, the most effective way to learn is to just do it. Pick a template program (e.g., Program 2.1) or snippet, make small, one-line-at-a-time changes, run, and see what you get. If an error pops up, do not be discouraged, follow debugging tips in Sect. 2.11. Otherwise, skip back and forth as you wish.

---

[6] As an alternative, try the more customizable all-in-one package WinPython, https://winpython.github.io/.

[7] https://www.python.org.

Python Tutorial

2

This chapter introduces the fundamental concepts of the Python language and will teach you to read and write Python code. Later chapters will build off the material covered here.

To help you learn, we recommend trying the examples yourself and testing any modifications you are curious about. You can use any of the IDEs mentioned in Chap. 1. If you are more comfortable starting out in a web browser environment without installing anything, you can try Jupyter Notebook on the web at https://jupyter.org/try.

2.1 Python as a Calculator

2.1.1 Arithmetic Operators

Python performs arithmetic with the familiar operators +, −, *, and /. Expressions can be grouped with parenthesis () to change the order of operations. For example,

```
In [1]: 1 + 2*3
```
```
Out[1]: 7
```
```
In [2]: (1 + 2)*3
```
```
Out[2]: 9
```

Powers are calculated with **. The following evaluates $2^3 \times 3^2 = 8 \times 9 = 72$.

```
In [3]: 2**3 * 3**2
```
```
Out[3]: 72
```

Note that ** precedes * in the order of operations. Perhaps less familiar are the floor and modulus operators // and %, which complete the set of arithmetic operators. The floor discards the remainder after division while the modulus only keeps the remainder.

J. Wang and A. Wang, *Introduction to Computation in Physical Sciences*, Synthesis Lectures on Computation and Analytics, https://doi.org/10.1007/978-3-031-17648-3_2

```
In [4]: 9 / 4
```

```
Out[4]: 2.25
```

```
In [5]: 9 // 4
```

```
Out[5]: 2
```

```
In [6]: 9 % 4
```

```
Out[6]: 1
```

In terms of order of operations: `**` has first priority; `*`, `/`, `//`, `%` are tied for second priority; and `+`, `-` are tied for last priority. Operators with tied priorities are evaluated left to right. Parenthesis can override these defaults and are useful in complicated expressions to avoid mistakes, even when not necessary. For instance, `(5 * 2**3)/2 + (1/2 + 3)` is far more readable than `5*2**3/2+1/2+3`.

2.1.2   Types of Numbers

Python treats integers, i.e. whole numbers, and floating-point numbers, i.e. numbers with decimal points, differently. They are represented by `int` and `float` data types, respectively. To check any value's type, use the `type()` function:

```
In [7]: type(1)
```

```
Out[7]: int
```

```
In [8]: type(2/3)
```

```
Out[8]: float
```

A value's type effects how it is stored in memory and what operations can be performed on it.[1]

Technicalities aside, precision is a major distinction. While integer arithmetic is exact and has its use cases, physical systems are continuous and floating-points are necessary. We must be aware of their limitations, however, shown in simple example $1 + 2/3 = 5/3$. In Python,

```
In [9]: 1 + 2/3
```

```
Out[9]: 1.6666666666666665
```

```
In [10]: 5/3
```

```
Out[10]: 1.6666666666666667
```

the results are not equal!

---

[1] For instance, it is perfectly valid to multiply two numbers, but what does it mean to multiply two words?

Python supports mixed arithmetic between different types of numbers. Expressions with integers and floating-points result in floating-points as we saw in `Out[9]`. Division always results in a floating-point, even in cases like 2 / 1.

Complex numbers have their own `complex` type. They are created using j as a suffix representing the imaginary part, e.g.

```
In [11]: type(1 + 1j)
```

```
Out[11]: complex
```

Arithmetic operators apply as usual, and they can be combined with `int` and `float` types.

```
In [12]: (1+2j)*(1-2j) + .5j
```

```
Out[12]: (5+0.5j)
```

2.2 Variables

It is helpful to store the result of a calculation in a variable for future use. Variable names also provide insight into the values they represent, which is especially useful in complicated calculations.

2.2.1 Assigning Variables

In Python, the equal sign = assigns a value to a variable.

```
In [13]: var1 = 1
```

The variable `var1` now has value 1. Note that no output is displayed when assigning variables. By default, the last line in a Jupyter Notebook cell is displayed if it is a value. Thus if output is desired, we can write an extra line of code, e.g.

```
In [14]: var2 = 2
         var2
```

```
Out[14]: 2
```

Code is executed sequentially line by line. Alternatively, we could use the `print()` function.

```
In [15]: var3 = 3/1
         print(var3)
```

```
         3.0
```

Once a variable has been assigned a value, it can be used in expressions with other variables and values. Variables can also be over-written. The following variable assignment may seem strange if we thought strictly mathematically:

```
In [16]:   var1 = var1 + var2**var3 + 1
           var1
```

```
Out[16]:   10.0
```

Python first evaluates the expression on the right, at which point var1 has value 1, then overwrites var1, assigning it the value 10.0. A single equal sign does not test for equality of two expressions; we will learn about logical operators in Sect. 2.6.1 that do.

2.2.2   Variable Names

Variable names can only contains letters, digits, and underscores, but cannot start with a digit. Variable names are case sensitive, e.g. var is different from VAR. Names should be meaningful so the code is more readable. For instance, if a variable represents an object's mass, it is reasonable to name it mass or m. Long variable names should not looklikethis, but should lookLikeThis or look_like_this; we will adopt the second convention in this book.

Python has certain built-in special words, like print; they are typically colored in IDEs. Variable names should not match these words, otherwise errors like the following will arise.

```
In [17]:   print = 5
           print(1)
```

```
---------------------------------------------------------------------------
TypeError                                           Traceback (most recent call last)
<ipython-input-17-5747096be880> in <module>
      1 print = 5
----> 2 print(1)

TypeError: 'int' object is not callable
```

The second line results in a TypeError because print is no longer a function (it now has type int), so we cannot call it with parenthesis like a usual function. We will learn more about errors and troubleshooting in Sect. 2.11. Default behavior can be restored by deleting the variable using del.

```
In [18]:   del print
           print(1)
```

```
1
```

2.3   Comments

While a computer may read and execute Python code with no trouble, it may not be easily understood by a human; this can be remedied by writing comments. Comments are text mixed with code that are ignored by Python. The hash symbol # denotes the start of a comment. The rest of the line starting from # is not executed as Python code. For example,

```
In [19]:  # Compute force, assuming constant mass and acceleration.
          m = 5     # mass in kg
          a = 2     # acceleration in m/s^2
          F = m*a   # Force in N
          print(F)
```

10

In this example, the comments are not necessary because it is simple and we chose meaningful variable names. However, as programs get more complicated, it is important to add comments to make the code easier to understand and troubleshoot.

2.4 Text Strings

Text strings, or simply strings, are used frequently. Some examples include loading/saving files, labeling figures, and displaying messages. This section covers the essentials of string manipulation.

2.4.1 Syntax

Strings are enclosed with either single quotes (`'a string'`) or double quotes (`"a string"`); the behavior is the same, but they cannot be mixed (`'not a valid string'`).

```
In [20]:  'This is string #1 with numbers and symbols. 1 + 2 = 3.'
```

```
Out[20]:  'This is string #1 with numbers and symbols. 1 + 2 = 3.'
```

```
In [21]:  "This is string #2. Double quotes work _just_ as well."
```

```
Out[21]:  "This is string #2. Double quotes work _just_ as well."
```

To use quotes within a string they must be prefixed by a \. For example,

```
In [22]:  "\"Here\'s a backlash \\\", they said."
```

```
Out[22]:  '"Here\'s a backlash \\", they said.'
```

The output may surprise you; the outer quotes are displayed as single quotes and the prefix \ is not displayed for the inner doubles quotes. The `print()` function parses a string and produces a more readable output.

```
In [23]:  print("\"Here\'s a backlash \\\", they said.")
```

```
"Here's a backlash \", they said.
```

Sometimes it is desirable to print over multiple lines. This is accomplished by \n within a string:

```
In [24]: print('Line 1.\nLine 2.')
```

```
Line 1.
Line 2.
```

Alternatively, we could use triple quotes (either single or double), which support multiline strings.

```
In [25]: """Line 1
Line 2
Line 3"""
```

```
Out[25]: 'Line 1\nLine 2\nLine 3'
```

There are a few other characters whose meaning is changed with a \ prefix. We will not be using them, but you can avoid this behavior by using a raw string, which is prefixed with an r. It is particularly useful for referring to directories like

```
In [26]: print(r'C:\some\directory\name')
```

```
C:\some\directory\name
```

What happens without the r?

2.4.2 String Operations and Indexing

Two strings can be combined with the + operator and a single string can be duplicated by multiplying with an integer.

```
In [27]: lower_letters = 'abc'
upper_letters = 'ABC'
lower_letters + upper_letters
```

```
Out[27]: 'abcABC'
```

```
In [28]: lower_letters*3 + 2*upper_letters
```

```
Out[28]: 'abcabcabcABCABC'
```

Individual characters in a string can be accessed with square brackets []. The first character has index 0, the second character has index 1, and so on.

```
In [29]: lower_letters[0]   # First character.
```

```
Out[29]: 'a'
```

```
In [30]: lower_letters[2]   # Third character.
```

```
Out[30]: 'c'
```

Negative indexes parse a string in reverse order, index −1 being the last character.

```
In [31]: lower_letters[-1]   # Last character.
```

```
Out[31]: 'c'
```

```
In [32]: lower_letters[-2]   # Second to last character.
```

```
Out[32]: 'b'
```

Substrings can be obtained by slicing from two indexes up to, but not including, the second index.

```
In [33]:  phrase = 'Hello, world!'
          phrase[0:5]  # First character to, but not including, 6th character.

Out[33]:  'Hello'

In [34]:  phrase[1:-1]  # Second character to, but not including, last character.

Out[34]:  'ello, world'
```

Omitting the first index slices from the beginning while omitting the second index slices until the end. Thus, `phrase[:i] + phrase[i:]` returns the entire phrase for any i and is equivalent to omitting both indexes, `phrase[:]`. Indexes can be out of bounds when slicing, but not when referring to a single element.

```
In [35]:  phrase[-100:100]

Out[35]:  'Hello, world!'

In [36]:  phrase[100]

          ---------------------------------------------------------------------
          IndexError                               Traceback (most recent call last)
          <ipython-input-36-c9fd83362889> in <module>
          ----> 1 phrase[100]

          IndexError: string index out of range
```

2.5   Functions

We will often repeat the same calculations, just with different parameters. Rather than typing out a small variant of the same code numerous times, we can define a function in Python to do the work for us. Functions have the added benefits of making code more readable (with proper name choices) and easier to troubleshoot. A good function name describes the essence of the operations it is performing. We have already encountered two built-in functions: `type()` and `print()`. We will learn how to define and use our own functions in this section.

2.5.1   Defining Custom Functions

A Python function can be similar to a mathematical function, say $f(x) = x^2$. It takes an argument x and returns its square, so $f(3) = 9$. Such a function is defined as follows.

```
In [37]:  def f(x):
              square = x**2
              return square

          f(3)  # Try our function.

Out[37]:  9
```

The term `def` is a Python keyword indicating a function is being defined, here named `f` with one argument x.[2] A colon indicates the function definition is complete. The rest of the code makes up the body of the function, and must be indented. The square of the argument is stored in a variable `square`, then `square` is returned using the `return` keyword. Reverting to normal indentation begins code that is no longer part of the function.

In general after a `return` statement, Python exits the function and does not execute the remainder of the code; this can be useful in complicated functions with conditional statements as we will see in Sect. 2.6.

For convenience, we can also define a single-line function called a `lambda` function,

```
In [38]: h = lambda x: x**2

         h(3)
```

Out[38]: 9

It is functionally equivalent to `f()` but simpler. Arguments are placed between the keyword `lambda` and the colon (separated by commas in case of multiple arguments), and the return value comes after the colon. A `lambda` function does not have a `return` statement.

Function names follow the same rules as variable names: they can only contain letters, digits, and underscores, but cannot start with a digit. Functions can have multiple arguments by separating arguments with commas, and can also have zero arguments.

```
In [39]: def complete_right_triangle(a, b):
             return (a**2 + b**2)**.5
             print('This line is never executed.')

         complete_right_triangle(3, 4)
```

Out[39]: 5.0

```
In [40]: def greet():
             print("Hello, there. Welcome to the world of Python!")

         greet()
```

```
Hello, there. Welcome to the world of Python!
```

Note that `greet()` does not have an explicit return value. By default, keyword `None` is returned, representing the absence of a value; it can only be displayed using `print()`.

```
In [41]: value = greet()
         value
```

```
Hello, there. Welcome to the world of Python!
```

```
In [42]: type(value)
```

Out[42]: NoneType

Functions can have a variable number of arguments; the `print()` function is one such example.

---

[2] Since functions are always used with a pair of parenthesis, to distinguish functions we will refer to them like `f()`, rather than just by name `f`.

```
In [43]:  print('Hello,', 'there.')      # 2 arguments.
          print()                         # No arguments (empty line).
          print(1, 2, 3, 'a', 'b', 'c')  # 6 arguments.
```

```
hello world
```

```
1 2 3 a b c
```

This can be done using *variable packing* and *unpacking* with syntax like `*args` or
`**kwargs`. Though we will not be heavily relying on this syntax, a basic example is
used in Sect. 3.2.1.

2.5.2   Variable Scope

Variables that are defined inside a function can only be used within that function; they are
called local variables. Variables defined outside functions can be used anywhere including
within functions; they are called global variables. Consider free-fall to illustrate this distinc-
tion, with vertical position $y(t) = -\frac{1}{2}gt^2 + v_0t + y_0$, where $g \approx 9.8\,\mathrm{m/s^2}$ and v_0, y_0 are
the initial speed and height.

```
In [44]:  g = 9.8   # Free-fall acceleration. Global variable.

          def get_freefall_ypos(t, y0, v0):
              y = -(g/2)*f(t) + v0*t + y0   # Local variable. Recall f(t) = t**2.
              return y

          print(get_freefall_ypos(2, 1, 0))
          print(y)
```

```
-18.6
---------------------------------------------------------------------------
NameError                                 Traceback (most recent call last)
<ipython-input-44-b344a94dcf5f> in <module>
      6
      7 print(get_freefall_ypos(2, 1, 0))
----> 8 print(y)

NameError: name 'y' is not defined
```

g is a global variable while y is a local variable within `get_freefall_ypos()`. The
function works as intended and has no issue using g, but an error in line 8 results from
using y outside its scope. Also note that we used a previously defined function `f()` within
`get_freefall_ypos()`. Any variables used in `f()` have their own local scope within
`f()`, and cannot be used anywhere else.

2.5.3   Default Argument Values and Keyword Arguments

Often times functions contain many arguments, but only a few are commonly changed. For example, if the most common initial condition is $y_0 = 1$ m and $v_0 = 0$ m/s, then we could specify these as default values by defining

```
In [45]:  def get_freefall_ypos(t, y0=1, v0=0):   # Default argument values.
              y = -(g/2)*f(t) + v0*t + y0
              return y
```

The new `get_freefall_ypos()` can be called with only one mandatory argument. If desired, one or more of the default arguments can be specified.

```
In [46]:  get_freefall_ypos(0)
```
```
Out[46]:  1.0
```
```
In [47]:  get_freefall_ypos(2, 0)   # Override one default argument, y0.
```
```
Out[47]:  -19.6
```

Rather than keeping track of the positions of each argument when calling a function, we can reference keyword arguments by their names, e.g.

```
In [48]:  get_freefall_ypos(t=2, y0=0, v0=0)   # Keyword arguments.
```
```
Out[48]:  -19.6
```

Keyword arguments must follow positional arguments, but otherwise can be specified in any order. Additionally, no argument can be assigned more than one value.

```
In [49]:  #get_freefall_ypos(t=2, 0, 0)    # Invalid; positional arg follows keyword arg.
          #get_freefall_ypos(2, 1, y0=1)   # Invalid; y0 assigned two values.
          get_freefall_ypos(y0=1, t=2, v0=0) # Valid.
          get_freefall_ypos(2, v0=0, y0=1)   # Valid.
```
```
Out[49]:  -18.6
```

2.6   Conditional Statements

So far all the code we have written has executed sequentially, line by line. Often times this is not sufficient to write a useful program. For instance, if we needed to compute the absolute value of x, we know mathematically

$$|x| = \begin{cases} x & \text{if } x \geq 0 \\ -x & \text{otherwise} \end{cases}.$$

That is, the returned value depends on the value of x itself. In Python, such a function relies on the `if` conditional statement and can be coded as follows.

```
In [50]:  def abs_val(x):
              if x >= 0:
                  return x
              else:
                  return -x
```

If $x < 0$, then the line `return x` is skipped over entirely.

2.6.1   Comparison and Logical Operators

Conditions that evaluate to true or false, represented in Python as keywords `True` and `False` (of type `bool`), are known as Boolean expressions. These conditions can be composed of six comparison operators, illustrated below.

```
In [51]:  1 == 2   # 1 is equal to 2? False.
          1 != 2   # 1 is not equal to 2? True.
          1 > 2    # 1 is greater than 2? False.
          1 < 2    # 1 is less than 2? True.
          1 >= 2   # 1 is greater than or equal to 2? False.
          1 <= 2   # 1 is less than or equal to 2? True.
```

```
Out[51]:  True
```

The operators `>`, `<`, `>=`, and `<=` only work for values of type `int` or `float`, while `==` and `!=` apply to any type. Note the equality operator `==` has an extra equal sign, distinguishing it from the assignment operator `=`. This can be confusing, and is important to remember to avoid errors.

Boolean expressions can be chained together to form more complex conditions using logical operators `and` and `or`. A third logical operator `not` returns the opposite of a Boolean expression. The following example shows these, using parenthesis to group Boolean expressions.

```
In [52]:  (1.2 > 0) and (1.2 < 1)    # False.
          (1.2 > 0) or (1.2 < 1)     # True.
          not (1.2 > 0)              # False.
```

```
Out[52]:  False
```

From highest to lowest priority, we have: (1) arithmetic operators (2) comparison operators (3) `not` (4) `and` (5) `or`. In practice, it is a good idea to use parenthesis to avoid errors.

2.6.2   `if` Statements

To execute code conditioned on a Boolean expression, use an `if` statement. For example, the following function returns `True` if `x` is positive.

```
In [53]:  def is_positive(x):
              if x > 0:
                  return True
```

The if keyword begins an if statement. It is followed by a Boolean condition and requires an ending colon. The body of the if statement, i.e. the code to execute if the condition is True, must be indented just like in a function. Python uses white space to distinguish blocks of code; each block indent is recommended to be 4 spaces wide for readability.

The is_positive() function has a limitation if $x \leq 0$, as None is returned. It would be better to return a Boolean value if we wanted to, for example, use the result of is_positive() in a Boolean expression. To cover all the cases when x is not positive, we can use an else part:

```
In [54]:  def is_positive(x):
              if x > 0:
                  return True
              else:    # x is not positive, i.e. x <= 0.
                  return False
```

The body of is_positive() could actually be coded in a single line with no if statement: return x > 0. However, this would only apply to functions with two possible conditions. We will often encounter functions that cannot partitioned this way, for instance the sign function

$$\text{sign}(x) = \begin{cases} 1 \text{ if } x > 0 \\ 0 \text{ if } x = 0 \\ -1 \text{ if } x < 0 \end{cases}.$$

To code sign(x), we can use elif parts, short for else if.

```
In [55]:  def sign(x):
              if x > 0:
                  return 1
              elif x == 0:
                  return 0
              elif x < 0:
                  return -1
```

Each condition is checked in order until a True condition is met. Only the True condition's code is executed; the remaining conditions are not checked. Thus, the order of elif parts matters if the conditions are not mutually exclusive. Any number of elif parts can be used, and the else part is optional. If no else part is present, it is possible that nothing in the if statement is executed.

if statements can be nested.

```
In [56]:  def is_positive_even(x):
              if is_positive(x):
                  if (x % 2) == 0:    # x is even.
                      return True
                  else:
                      return False
              else:
                  return False
```

Many nested if statements should be avoided for readability if there is a simpler alternative. The above can be simplified as follows.

```
In [57]:  def is_positive_even(x):
              return (x % 2 == 0) and (x > 0)
```

2.7 Loops

This section covers `while` and `for` loops to repeatedly execute code, and how to break out of them early if needed.

2.7.1 `while` Loops

Unlike `if` statements that execute code a single time if a condition is `True`, `while` loops execute code over and over while a condition is `True`. Their syntax is similar, shown in the following function that computes factorials for integer n:

$$n! = \begin{cases} 1 & \text{if } n = 0 \\ n \times (n-1) \times \cdots \times 1 & \text{if } n > 0 \end{cases}.$$

```
In [58]:  def factorial(n):
              if n == 0:
                  return 1

              prod = 1
              while n > 0:
                  prod = prod * n
                  n = n - 1
              return prod
```

The familiar `if` statement treats the $n = 0$ case separately. The `while` loop executes as long as $n > 0$ and eventually terminates because at each iteration, n is decremented by one. Be careful, as it is possible to enter an infinite loop if the condition checked will never be false.

2.7.2 `for` Loops

It is often useful to loop through a known sequence. The `range()` function is particularly useful in `for` loops to iterate over sequences of integers. A simple `for` loop iterating over $i = 0, 1, 2$ is as follows.

pagebreak

```
In [59]:  for i in range(0, 3):
              print(i)
```

```
Out[59]:  0
          1
          2
```

The syntax reads similar to its English meaning. `for` and `in` are Python keywords, and `i` is a variable representing an element in the sequence `range(0, 3)`. In the first iteration of the `for` loop, `i` is assigned the first element of the sequence. The next iteration assigns `i` to be the next element of the sequence and so on, until all elements of the sequence are looped through. The above example simply prints each element.

The `range()` function arguments behave like string slicing. `range(a, b)` represents the sequence of integers starting from `a` up to, but not including, `b`. Omitting an argument, e.g. `range(n)`, assumes the sequence begins from zero up to, but not including, `n`.

Strings can be viewed as a sequence of letters, and thus can also be used in a `for` loop.

```
In [60]:  for letter in 'xyz':
              print(letter)

Out[60]:  x
          y
          z
```

We used a relevant variable name `letter`, although any variable name would do. We will learn about other sequences that can be looped through, e.g. lists, in Sects. 2.7.3 and 2.9.

2.7.3 **break** and **continue** Statements

Loops can be terminated early using a `break` statement. The next iteration of a loop can be forced without executing the remainder of the code using a `continue` statement. The following code checks for commas in a phrase, and exits the loop after a comma is found.

```
In [61]:  for char in 'Hello, world!':
              if char == ',':
                  print('No commas allowed!')
                  break   # Break out of loop.
              else:
                  continue   # Immediately continue next iteration of loop.
              print('This line is never executed.')

          No commas allowed!
```

A `break` statement breaks out of the current loop. This is an important distinction with nested loops:

```
In [62]:  for vowel in 'aeiou':
              for char in 'Hello, world!':
                  if char == vowel:
                      break   # Break out of (current) loop.
              print(vowel)   # This line is always executed.

          a
          e
          i
          o
          u
```

If we wanted to break out of all loops, we should insert the loops into a function and `return` out of the function.

2.8   Lists

We've encountered two types of sequences so far: the `range()` function for numeric sequences and strings. Lists are common sequences used to group together values.

2.8.1   Syntax

Lists are denoted by square brackets, with elements separated by commas.

```
In [63]:  nums = [1, 2, 3]            # nums is a list of numbers.
          letts = ['a', 'b', 'c']  # letts is a list of letters.
          nums_and_letts = [1, 'a', 2, 'b']
          nums_and_letts
```

```
Out[63]:  [1, 'a', 2, 'b']
```

List elements can be any data type, including nested lists. The `len()` function returns the number of elements. A list can contain zero elements.

```
In [64]:  empty_list = []
          diverse_list = [[1, 2, 3], 4, '5', 'xyz']
          len(diverse_list)
```

```
Out[64]:  4
```

2.8.2   List Operations and Indexing

Lists operations are similar to strings; they can be combined with + and duplicated by integer multiplication.

```
In [65]:  nums + letts
```

```
Out[65]:  [1, 2, 3, 'a', 'b', 'c']
```

```
In [66]:  diverse_list*2
```

```
Out[66]:  [[1, 2, 3], 4, '5', 'xyz', [1, 2, 3], 4, '5', 'xyz']
```

Indexing and slicing return sublists, with syntax just like in strings.

```
In [67]:  diverse_list[0][-1]   # Last element of diverse_list's first element.
```

```
Out[67]:  3
```

```
In [68]:  diverse_list[1:-1]  # Second element to, but not including, last element.
```

```
Out[68]:  [4, '5']
```

```
In [69]:  diverse_list[2:]   # Third element to the end.
```

```
Out[69]:  ['5', 'xyz']
```

Elements of lists, and even slices, can be updated using the assignment operator.

```
In [70]:  nums[-1] = 4  # Reassign last element.
          nums
```

```
Out[70]:  [1, 2, 4]
```

```
In  [71]:  nums[1:3] = [3, 9]  # Reassign second and third element.
           nums
```

```
Out[71]:  [1, 3, 9]
```

Lists are mutable data types, meaning their contents can be changed. In contrast, strings are immutable data types and do not support reassignment of elements.

```
In  [72]:  'abc'[0] = 'A'  # Try reassigning first element of a string.
```

```
           --------------------------------------------------------------------
           TypeError                                 Traceback (most recent call last)
           <ipython-input-72-94a5227817b6> in <module>
           ----> 1 'abc'[0] = 'A'  # Try reassigning first element of a string.

           TypeError: 'str' object does not support item assignment
```

To check if a particular value is in a list, we can use the `in` operator.

```
In  [73]:  9 in nums
```

```
Out[73]:  True
```

```
In  [74]:  3 not in nums
```

```
Out[74]:  False
```

2.8.3 List Methods

A method is like a function, only it belongs to a specific object and is called using dot syntax. List methods belong to objects with data type `list`. For example, the `append()` method adds a new element to the end of a list, and can be used as follows.

```
In  [75]:  nums = [1, 2, 3]
           nums.append(4)  # Append 4 to the end of the list nums.
           nums
```

```
Out[75]:  [1, 2, 3, 4]
```

In conjunction with a `for` loop, a list of integers can be easily generated.

```
In  [76]:  ints = []
           for i in range(100):
               ints.append(i)
```

The `append()` method is useful in many calculations to store results. For example, using our function to compute free-fall positions, we can store a trajectory $y(t)$ for $0 \leq t < 1$ in a list.

```
In  [77]:  y = []
           t = 0
           while t < 1:
               y.append(get_freefall_ypos(t, y0=0, v0=0))
               t = t + .01
```

To insert an element at a specific position, use the `insert()` method.

```
In [78]: i = 2
         x = 5
         nums.insert(i, x)   # Insert x before the i-th index of nums.
         nums
```

Out[78]: [1, 2, 5, 3, 4]

In contrast to appending, `pop()` removes an element of the list and returns its value. An optional argument specifies the index of the element to remove and return; by default, it is the last element.

```
In [79]: nums.pop()   # Remove and return last element of nums.
```

Out[79]: 4

```
In [80]: nums
```

Out[80]: [1, 2, 5, 3]

To remove a particular value (rather than remove by index), the `remove()` method should be used. Only the first instance of the argument value is removed. If there is no such value, an error occurs.

```
In [81]: nums.remove(2)   # Remove the first element with value 2.
         nums
```

Out[81]: [1, 5, 3]

To return the index of the value instead, use the `index()` method.

Lists can be sorted in either ascending or descending order with the `sort()` method.

```
In [82]: nums.sort()   # Sort nums from smallest to largest.
         nums
```

Out[82]: [1, 3, 5]

```
In [83]: nums.sort(reverse=True)   # Sort nums from largest to smallest.
         nums
```

Out[83]: [5, 3, 1]

The methods mentioned with the exception of `pop()` do not have return values. Therefore code like `nums = nums.sort()` may not perform what you expect it to. The list is sorted, but nothing is returned; thus after reassignment, `nums` would have the value `None`.

2.9   Tuples, Sets, and Dictionaries

So far we have encountered the following built-in data types: integers, floating-point numbers, complex numbers, strings, Booleans, range, and lists. There are three more we will be using, all introduced in this section.

2.9.1   Tuples

Tuples are similar to lists. They are denoted by parenthesis, with elements separated by commas.

```
In [84]: tup = (1, 'a', [2, 'b'], (3, 'c'))
         len(tup)
```

Out[84]: 4

Tuples of length one must have a comma like $t = (1,)$ to differentiate from an arithmetic expression. Indexing works just like in lists.

Unlike lists, tuples are immutable. Their elements cannot be reassigned, although they can contain mutable elements whose subelements can be reassigned, for instance

```
In [85]: tup[2][0] = 222
         tup
```

Out[85]: (1, 'a', [222, 'b'], (3, 'c'))

Tuples can be used for multiple assignment. The following assigns $a = 3, b = 4$, and $c = 5$ as if each assignment was made on a separate line.

```
In [86]: (a, b, c) = (3, 4, 5)   # Multiple assignment.
         a**2 + b**2 == c**2
```

Out[86]: True

This is technically known as sequence packing and unpacking, and requires an equal number of elements on both sides. When defining tuples, the parenthesis are optional (except for empty tuples), so sometimes you will see code like the following.

```
In [87]: a, b = b, a   # Swap values of a and b.
```

2.9.2   Sets

The sequences encountered so far have all been ordered. Python sets are unordered much like mathematical sets, e.g. a set representing the outcomes of a coin toss $S = \{H, T\} = \{T, H\}$. Because sets are unordered, their elements cannot be indexed. However, a for loop can iterate through all its elements. Sets are denoted with curly braces.

```
In [88]: {'H', 'T'} == {'T', 'H'}   # Are these sets equal?
```

Out[88]: True

```
In [89]: ['H', 'T'] == ['T', 'H']   # Are these lists equal?
```

Out[89]: False

Sets have no duplicate elements, and support set operations like union and intersection.

```
In [90]: s1 = {8, 1, 3, 3, 5, 3}
         s1
```

```
Out[90]: {1, 3, 5, 8}
```

```
In [91]: s2 = {2, 3, 4, 5}
         s1 & s2  # Set intersection.
```

```
Out[91]: {3, 5}
```

```
In [92]: s1 | s2  # Set union.
```

```
Out[92]: {1, 2, 3, 4, 5, 8}
```

An easy way to remove duplicates from a sequence is to convert to a set using the `set()` function.

```
In [93]: nums = [1, 1, 2, 3, 4, 4]
         unique_nums = set(nums)  # Convert nums to a set.
         unique_nums
```

```
Out[93]: {1, 2, 3, 4}
```

Similarly, `list()` and `tuple()` convert a sequence to a list and tuple, respectively. These functions with no argument return empty sequences. In particular, `set()` is the only way to create an empty set; a pair of curly braces returns an empty dictionary.

2.9.3 Dictionaries

Python dictionaries store key and value pairs, similar to word dictionaries. For example, a Python dictionary mimicking a simple word dictionary is as follows.

```
In [94]: word_dict = {
             'programming': 'The action or process of writing computer programs.',
             'Python': 'A high-level general-purpose programming language.'
         }

         word_dict['Python']  # Return value associated with the key 'Python'.
```

```
Out[94]: 'A high-level general-purpose programming language.'
```

This dictionary's keys are words, and the associated value of each key is a definition. To use the dictionary, we can reference by keys to return the associated values. This dictionary definition also introduced code that spans multiple lines but acts as a single line; we saw similar syntax in multiline strings (`In [25]`). Python gracefully handles so called line continuations inside parentheses `()`, brackets `[]`, and braces `{}`; we added optional indentation to make the line continuations clear.[3]

In general the syntax of a Python dictionary are `key: value` pairs separated by commas inside curly braces. The keys must be immutable and should be unique, while values are flexible. Elements can be added after a dictionary `d` is declared with syntax `d[key] = value`.

---

[3] For expressions, conditions, or many argument functions that would result in a long line, an outer pair of parenthesis with indentation can be used for line continuation.

As another example, suppose we perform a simulation where we flip two coins and record the outcome many times. The sample space is the set {HH, HT, TH, TT}. A dictionary is a natural data type to record the frequency of each outcome.

```
In [95]: results = {'HH': 24, 'HT': 19, 'TH': 26, 'TT': 31}
```

A for loop can iterate over key and value pairs using a dictionary's items() method.

```
In [96]: for key, val in results.items():
             print(key, ':', val)
```

```
Out[96]: HH : 24
         HT : 19
         TH : 26
         TT : 31
```

Keys and values are obtained individually from the keys() and values() methods, which can be converted into other sequences with the list(), tuple(), and set() functions.

2.10   Modules and Packages

We will often use the same code, e.g. common functions, in programs we write. Rather than copying and pasting into every program, we can import a module. Think of a module as a single program that contains several definitions within. As an example, we will frequently use the math module, which is in the standard library of modules that comes with Python. To access its functions and mathematical constants like $\pi = 3.14159...$ and $e = 2.71828...$, we must first import the module and then use dot syntax. For example, the following computes $\cos(\pi)$.

```
In [97]: import math     # Import math module.

         math.cos(math.pi)   # Compute cos(π).
```

```
Out[97]: -1
```

(We typeset math in the comments using LaTeX rather than something like `cos(pi) = -1`. Only the latter is possible when writing Python code.) The math module contains definitions for cos() and pi. Note that math.pi is not a function, hence no parenthesis, while math.cos() is.

We will use the random module, also in the standard library, for all things random number generation related. The following code generates a list of ten random integers between 1 and 5.

```
In [98]: import random

         random_ints = []
         for i in range(10):
             random_ints.append(random.randint(1, 5))
```

We will also frequently use modules (and packages, see below and Chap. 3) not in the standard library. These require additional installation; see Sect. 3.2.2 for an example.

Modules generally group together a similar set of definitions. As they become more complex, it can be convenient to structure a collection of modules into a package. Submodules of the package are accessed using dot syntax. We will frequently use the `pyplot` module for plotting, which is in the `matplotlib` package. To import it (assuming it is installed), type

```
In [99]: import matplotlib.pyplot
```

To avoid name conflicts with variables, functions, or other modules, we can import a module as a particular name. This is also convenient for nested modules or long module names. For example, it is conventional to import `pyplot` using

```
In [100]: import matplotlib.pyplot as plt
```

Now rather than typing `matplotlib.pyplot.function()` every time we want to use `function()`, we can type `plt.function()`. A basic plot is created as follows.

```
In [101]: %matplotlib inline
          plt.plot(random_ints)
          plt.xlabel('trial')
          plt.ylabel('random integer')
```

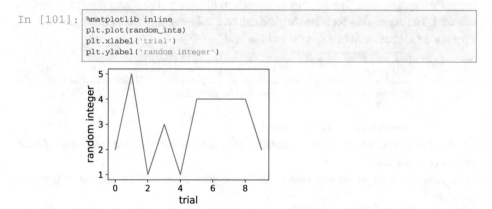

The first line is necessary in Jupyter Notebook to output the figure directly after the cell. If you are running your program as a script, you may need the line `plt.show()` at the end to display your figure. We will learn more about plotting and visualization in Sect. 3.2.1.

2.11 Error Handling and Troubleshooting

Errors that prevent a program from executing are common among beginner and experienced programmers alike and are often referred to as bugs. This section will discuss a few common errors, how to interpret error messages and use them to fix code, and how to write code that can handle errors without preventing the program from crashing over minor bugs.

2.11.1 Syntax Errors

Syntax errors occur when code is not understandable to the computer, and thus unable to be executed. They are common, but generally straightforward to troubleshoot due to the resulting error message. The process of debugging, namely tracking down and fixing errors, is important and not to be afraid of, and every coder from the novice to the expert makes mistakes.

For example, mathematically the expression $5x + 2$ is valid, but in Python

```
In [102]:   x = 5
            5x + 2

              File "<ipython-input-102-53ff5f61ab1a>", line 2
                5x + 2
                 ^
            SyntaxError: invalid syntax
```

results in a syntax error. The error message tells us the line and character the syntax error is detected. The expression $5x$ is invalid, and can be fixed by adding an explicit multiplication operator $5*x$. A more subtle error is as follows.

```
In [103]:   if x > 5
                print(x)

              File "<ipython-input-103-b44ee23f11de>", line 1
                if x > 5
                       ^
            SyntaxError: invalid syntax
```

Every if statement needs a colon to terminate the condition in question; no colon is found, so this code is unable to execute.

If there are multiple syntax errors, the first detected one will be located in the error message.

```
In [104]:   if 5x > 5    # Errors: 5x is invalid, missing colon.
                print(5*x larger than 5)  # Error: Space separated expressions.

              File "<ipython-input-104-68cae071cbe3>", line 1
                if 5x > 5    # Errors: 5x is invalid, missing colon.
                    ^
            SyntaxError: invalid syntax
```

2.11.2 Exceptions

Errors not related to syntax are known as exceptions. They can be more difficult to troubleshoot because the code looks correct at a glance. There are many kinds of exceptions, all with a different error message. For example,

```
In [105]: x = 5
          1/(x-5)
```

```
---------------------------------------------------------------------
ZeroDivisionError                          Traceback (most recent call last)
<ipython-input-105-7dff629e6398> in <module>
      1 x = 5
----> 2 1/(x-5)

ZeroDivisionError: division by zero
```

indicates there is a division by zero error detected on line 2. More complicated errors have longer tracebacks that propagate the error, making it easier to troubleshoot. As an example, recall our abs_val() function which computes absolute values. Suppose we had a list of values and tried the following to compute all their absolute values:

```
In [106]: abs_val([-1, 0, 1])
```

```
---------------------------------------------------------------------
TypeError                                  Traceback (most recent call last)
<ipython-input--106-68ac0f640e22> in <module>
----> 1 abs_val([-1, 0, 1])

<ipython-input--50-ecc5f98eaf28> in abs_val(x)
      1 def abs_val(x):
----> 2     if x >= 0:
      3         return x
      4     else:
      5         return -x

TypeError: '>=' not supported between instances of 'list' and 'int'
```

The most recent call results in the TypeError because it does not make sense to use the comparison operator >= between a list and an integer. The error is traced back to the previous function call erroneously using a list as an argument to abs_val().

Two other common exceptions are IndexError, from accessing an index in a sequence that does not exist (e.g. In [36]), and NameError, from calling a variable that does not exist (e.g. In [44]). As code gets more complicated, the number of errors will increase and tracebacks become longer. For this reason, it is a good idea to test small code segments as you program.

2.11.3 Handling Exceptions

Suppose we were interested in converting strings like '123' to integers; this can be done using the int() function. However, if the string is not integer-like, e.g. 'abc', a ValueError occurs. To handle this exception smoothly, we can use a try and except statement, whose syntax is similar to if (and elif) statements.

```
In [107]:  string = 'not an integer'
           try:
               x = int(string)
           except ValueError:   # In case string is not integer-like.
               x = None
           except TypeError:   # In case string is not actually a string.
               x = None

           print(x)
```

None

First, the code tries to assign x to int(string). If successful, the except statements
are ignored. If there is a ValueError or TypeError, x is assigned None rather than
crashing the program.

try and except statements are especially useful when repeating a similar calculation
numerous times, e.g. as a function of some parameter(s). Perhaps certain (unknown) values
of the parameter have convergence issues or division by very small numbers leading to
errors. Rather than restarting the simulation with an additional constraint on the parameters,
we could handle exceptions smoothly and store the problematic parameters in a list for later
inspection.

2.11.4 Errors with No Error Message

Sometimes code executes fine, but something about the results seems off. The reason can
vary from a forgotten minus sign, a serious logic error, or even no errors at all! These are
the hardest errors to troubleshoot, especially when you are not sure what to expect in your
results. How can we be sure our program is error free, and the result is valid? Unfortunately
we can make no guarantees, but there are tools and strategies to make us confident in our
code.

We again emphasize troubleshooting in small chunks that are simple enough to be easily
tested, for instance functions written with a purpose in mind. Functions have the added
benefit that once extensively tested are unlikely to be the source of error in the program. For
example, recall our function that computes absolute values defined in In [50]. A simple
test verifying it works for negative and positive numbers and zero is as follows.

```
In [108]:  for x in [-1, 0, 1, -math.pi]:
               print(abs(x))
```

    1
    0
    1
    3.141592653589793

If our program only uses it to calculate absolute values of real numbers, then we have shown
it is robust. It can still fail for other data types; see Exercise 2.14 for a generalization.

Even if all the code segments work as intended, they may interact unexpectedly. Troubleshooting an entire program is more difficult, but can be managed by inserting probes
in the form of print() statements at key locations. These enable snapshots of salient

variable values, when certain conditionals are executed, or when a function is called. Even better, many IDEs (e.g. Spyder, see Sect. 1.2) have integrated debuggers that allow the user to set breakpoints. The user can then step through the program, stopping at each breakpoint to inspect variables and the flow of the program.

2.12   A Full Program

This section will combine elements of the tutorial into a single, self-contained program that solves the following question:

A ball is thrown from height $y_0 = 1$ m at speed $v_0 = 3$ m/s. What angle should it be thrown to maximize the horizontal distance until it hits the ground ($y = 0$ m), in the absence of air resistance?

While this question can be solved analytically, it cannot in the more realistic scenario with air resistance. Meanwhile, the programmatic approach is easily modified as we will see in Sect. 5.4.

2.12.1 Programmatic Solution

We can construct a crude algorithm relying only on the relations between position, velocity, and acceleration. Consider one dimension for now: over a time period Δt, the average velocity and acceleration are $\bar{v} = \Delta y / \Delta t$ and $\bar{a} = \Delta v / \Delta t$. Using $\Delta y = y(t + \Delta t) - y(t)$, the first relation can be rearranged to

$$y(t + \Delta t) = y(t) + \bar{v} \Delta t \qquad (2.1)$$

which says at some time Δt later, the position will be incremented by $\bar{v} \Delta t$. However given initial conditions y_0 and v_0, we only know the velocity at time t and not the average velocity over the time interval $[t, t + \Delta t]$. We can make a crude assumption that for small time intervals, the average velocity is approximately the velocity at the start of the interval $v(t)$. Then we have the update rule

$$y(t + \Delta t) \approx y(t) + v(t) \Delta t. \qquad (2.2)$$

The same argument updates velocities as

$$v(t + \Delta t) \approx v(t) + a(t) \Delta t. \qquad (2.3)$$

This algorithm is known as Euler's method and is only exact when the underlying system behaves linearly. More accurate methods will be discussed in Sect. 5.5.3.

As an example to illustrate these rules, we will revisit our `get_freefall_ypos()` function in `In  [45]` and implement it using Euler's method.

```
In [109]:  def euler_freefall_ypos(tf, dt=.01, y0=0, v0=0):
               t, y, v = 0, y0, v0
               while t < tf:
                   t = t + dt
                   y = y + v*dt        # Eq.(2.2)
                   v = v + (-g)*dt     # Eq.(2.3)
               return y

           euler_freefall_ypos(2, dt=.01)
```

Out[109]: -19.502000000000027

Note the deviation from the (quadratic) analytic solution of -19.6. The error can be reduced by lowering the timestep at the cost of longer simulation time.

```
In [110]:  euler_freefall_ypos(2, dt=.001)
```

Out[110]: -19.60980000000014

We are now ready to implement a programmatic solution. In Program 2.1, we use Euler's method to compute two dimensional projectile motion trajectories and record the horizontal range as a function of throw angle. Program 2.1 is divided into five labeled sections and ends by printing the range-maximizing angle.

Program 2.1 A full program to compute range-maximizing throw angles.

```
1  ### 1) Begin a program with import statements.
   import math
3
   ### 2) Define functions.
5  def projectile_range(x0, y0, v0, theta, dt=.01):
       """Compute range of projectile launched from (x0, y0) at speed v0 and
7      angle theta relative to the horizontal (in radians)."""
       t, x, y = 0, x0, y0
9      vx, vy = v0*math.cos(theta), v0*math.sin(theta)
       while y >= 0:
11         # Update using Euler's method. (There is no horizontal acceleration.)
           t = t + dt
13         x = x + dt*vx
           y = y + dt*vy
15         vy = vy - dt*g
       return x - x0
17
   def deg_to_rad(angle):
19     """Convert an angle from degrees to radians."""
       return angle * (math.pi/180)
21
   ### 3) Define and initialize parameters.
23 g = 9.8                    # Free-fall acceleration.
   x0, y0, v0 = 0, 1, 3       # Initial positions (m) and speed (m/s).
25 dtheta = .1                # Angle resolution (deg).
   angles, ranges = [], []    # Lists to store angles and associated ranges.
27 for i in range(int(90/dtheta)):
       angles.append(i*dtheta)    # Append 0, dtheta, 2*dtheta, ..., up to 90 deg.
29
   ### 4) Run simulations.
31 for angle in angles:
       theta = deg_to_rad(angle)
33     r = projectile_range(x0, y0, v0, theta)
       ranges.append(r)
```

```
35
   ### 5) Simulation analysis.
37 max_index = ranges.index(max(ranges))
   max_angle = angles[max_index]
39 print(max_angle)   # 29.0
```

2.12.2 Analytic Solution

The kinematic equations are

$$x(t) = v_{0x}t + x_0$$
$$y(t) = -\frac{1}{2}gt^2 + v_{0y}t + y_0$$

where the components of velocity are related to the launch angle θ by $v_{0x} = v_0 \cos\theta$, $v_{0y} = v_0 \sin\theta$. The projectile reaches the ground at time[4]

$$t_* = \frac{1}{g}\left(v_{0y} + \sqrt{v_{0y}^2 + 2gy_0}\right),$$

corresponding to a traversed range of

$$R(\theta) = x(t_*) - x(0) = \frac{v_{0x}}{g}\left(v_{0y} + \sqrt{v_{0y}^2 + 2gy_0}\right).$$

Setting $dR/d\theta = 0$ and performing some algebra results in the range-maximizing angle

$$\theta_* = \cos^{-1}\sqrt{\frac{2gy_0 + v_0^2}{2gy_0 + 2v_0^2}}.$$

As we would expect, $0° \le \theta_* \le 90°$. At finite speeds, the optimal angle is only $45°$ when $y_0 = 0$. For $y_0 > 0$, $\theta_* < 45°$ while for $y_0 < 0$, $\theta_* > 45°$. With $y_0 = 1$ m and $v_0 = 3$ m/s, $\theta_* \approx 29°$.

With air resistance, the logic behind Eqs. (2.2) and (2.3) remains the same. In general, the update rules take the form

$$y(t) \approx y(t) + f(y, v, t)\Delta t$$
$$v(t) \approx v(t) + g(y, v, t)\Delta t$$

(2.4)

where $f(\cdot)$ and $g(\cdot)$ can be more complicated functions. The analytic solutions, on the other hand, become much more involved.

---

[4] When $y_0 < -v_{0y}^2/2g$, the initial height is too low relative to the velocity that the height $y = 0$ cannot be reached, even with a vertical $\theta = 90°$ throw.

2.13   Classes

Sometimes it is useful to explicitly bundle data and functions together. As you write more complex programs and notice you have a group of functions that repeatedly use the same arguments, e.g. the same set of initial conditions or parameters, then it can be useful to bundle them together in a Python `class`. This paradigm is known as *object oriented programming*. We illustrate with a basic example—a more involved one is deferred to Sect. 9.1.2.

Suppose we have a soccer ball and wish to track its position as we kick it forward. One way to do this is to define initial conditions, compute the range of a kick given launch parameters, update the conditions, and give another kick with the new conditions:

```
In [111]:   x = 0
            x = x + projectile_range(x, y0=0, v0=1, theta=.5)
            x = x + projectile_range(x, y0=0, v0=2, theta=.5)
```

Supposing we want to throw the ball next, we must continue passing and updating the same *x* parameter:

```
In [112]:   height = 1
            x = x + projectile_range(x2, y0=height, v0=1, theta=.4)
            x
```

```
Out[112]:   0.9256492548033407
```

The object oriented approach defines a `Ball` object using Python keyword `class`:

```
In [113]:   class Ball:
                def __init__(self, x=0, height=1):
                    self.x, self.height = x, height

                def kick(self, v0, theta, dt=.01):
                    self.x += projectile_range(self.x, 0, v0, theta, dt)

                def throw(self, v0, theta, dt=.01):
                    self.x += projectile_range(self.x, self.height, v0, theta, dt)
```

The `Ball` class has three *methods*, which can be thought of as functions that belong to the `Ball` class. The `__init__()` method is special—whenever a `Ball` object is created, this method is called and it assigns the `x` and `height` attributes to the object itself. The positional `self` argument can be accessed anywhere within the `Ball` class, meaning every other method has access to the ball's position and throw height. This is precisely how functionality and data are bundled together.

We can create instances of the `Ball` object using syntax as if it were a function:

```
In [114]:   ball = Ball()
            ball.x, ball.height
```

```
Out[114]:   (0, 1)
```

We see that its initial position and height have been properly initialized. Next, we can call the `kick()` and `throw()` methods, which also have the `self` argument. This allows them to access the position, no matter how many times its been kicked or thrown, and compute the resultant position.

```
In [115]: ball.kick(v0=1, theta=.5)
          ball.kick(v0=2, theta=.5)
          ball.throw(v0=1, theta=.4)
          ball.x
```

Out[115]: 0.9256492548033407

After each kick and throw, the ball's x position is updated and remembered for the next kick or throw. As programs get more complex, this allows us to group functionality together without passing the same parameters in multiple different places.

As a final note, the `Ball` class is not the first class we have encountered. Indeed Python lists are themselves classes!

```
In [116]: x = []
          print(type(x))
```

Out[116]: <class 'list'>

Their functionality was tied to methods like append(), which were called with the same dot syntax. We will use many modules and packages that have a large chunk of functionality implemented as classes. This allows for convenient usage, as we can access all relevant functions with simple dot syntax.

2.14 Exercises

2.1. A generalized absolute value function. We implemented a basic version of $f(x) = |x|$ in Sect. 2.6 that acts on a scalar number x. Extend the function so that it can also operate on a list of numbers as input and return a list of absolute values as output. Print the elements and their index where an absolute value cannot be calculated, and remove them from the final output. The following are expected behavior:

```
In [117]: abs_val(-2)
```

Out[117]: 2

```
In [118]: abs_val([-1, 'not_a_number', -3])
```

Out[118]: Warning: Dropping input index 1, cannot calculate abs of
          'not_a_number' [1, 3]

2.2. Random Python practice.

(a) Generate a list of 100 random integers between 1 and 100. Calculate the mean, median, and mode of that list.

(b) Store the number of times each integer appears in a dictionary.

(c) Assign each integer into one of ten groups, representing equal width bins [1, 10], [11, 20], ..., [91, 100]. Continue generating random integers until at least two groups contain the same number of integers.

2.3. Debugging practice. Let us take a closer look at Program 2.1. Although the program runs and appears to give reasonable output, closer inspection shows that it does not match analytic results exactly (within specified precision).

(a) Explain why you would expect the range-maximizing angle in Program 2.1 to be accurate to one decimal place.

(b) With the same initial conditions as the text, $y_0 = 1\,\text{m}$ and $v_0 = 3\,\text{m/s}$, use the analytic solution (Sect. 2.12.2) to calculate the exact range-maximizing angle, accurate to one decimal place.

(c) The angles in the previous two parts are close, but do not match exactly. Modify Program 2.1 to plot the range as a function of angle and describe what is unusual about the plot. Investigate a single range calculation to determine what the issue is. Does decreasing the timestep to $\texttt{dt} = .001$ help? Check graphically and in comparison to the analytic solution.

(d) Implement a solution to improve the accuracy of the range calculation.

2.4. All about primes.

(a) Write a function to determine if n is a prime number. You can assume n is a positive integer. Is 9,521,928,547 prime?

(b) What is the 10^4-th prime number?

(c) How many numbers less than 10^7 are prime?

(d) Plot the ratio n_{prime}/n for $2 < n < 10^3$, where n_{prime} is the number of primes $\leq n$. Describe your results. What would you expect as n continues to increase?

2.5. Collatz sequences. Beginning with a positive integer a_0, consider the Collatz sequence generated by iteratively applying

$$a_{i+1} = \begin{cases} a_i/2 & \text{if } a_i \text{ is even} \\ 3a_i + 1 & \text{if } a_i \text{ is odd} \end{cases}, \quad i = 0, 1, 2, \ldots.$$

It is conjectured that all sequences eventually reach 1. For example, starting from $a_0 = 10$ yields the sequence $[10, 5, 16, 8, 4, 2, 1]$ of length $L = 7$, i.e. it contains 7 elements until 1 is reached. In this problem, we will consider all sequences starting from $a_0 = 2, 3, \ldots, 10^6$.

(a) Which a_0 yields the longest sequence? What is its length?

(b) Plot L as a function of a_0 and the distribution of L. Restrict the $L(a_0)$ plot to $a_0 < 10^4$.

(c) Since every odd integer will become even after one iteration while half the even integers remain even, the number of odd terms in a sequence $n_{odd} \leq n_{even}$ (aside for $a_0 = 1$). Plot the ratio $r = n_{odd}/n_{even}$ for each a_0 and the distribution of ratios, restricting the $r(a_0)$ plot to $2 < a_0 < 10^4$. Show that r is bounded by $\log 2/\log 3$ for $a_0 > 2$.

2.6. Repeating decimals. Dividing two integers results in a rational number n/m, whose decimal representation either repeats (e.g. $1/11 = 0.\overline{09}$) or terminates (e.g. $1/2 = 0.5$). Write a function that computes the cycle length of n/m, defining a terminating decimal to have cycle length $L = 0$. Plot $L(d)$ for fractions of the form $1/d, d = 2, 3, \ldots, 10^3$, and the distribution of lengths. Comment on any patterns you observe.

Interactive Computing and Visualization

3

As discussed earlier in Chap. 1, computational environments have evolved from central computing in the early days to distributed computing, and from text-based terminal interfaces to graphical and web-based development environments at present. The latter offers new and more interesting possibilities. In this chapter, we discuss two such possibilities: interactive computing and visualization. The advantage of interactive computing is multi-fold: a program can be composed of simpler code blocks, each individually controlled with immediate output displayed after execution; models can be tweaked and parameters changed freely to see the results instantaneously to test what-if scenarios as if turning knobs in a virtual experiment. Effective visualization has always been an important aspect in computer modeling, but traditionally is treated as a separate process from computation. This, too, can be integrated seamlessly in the web-based computing environments including in-situ data visualization and real-time animation. In the following sections, we describe interactive elements such as widgets, visualization and animation techniques, and useful libraries for our purpose.

3.1 Interactive Computing with Python Widgets

We can use widgets for interactive computing in standard Python via `matplotlib` or in `Jupyter` via `ipywidgets`. Typical widgets include sliders, buttons, and menus. The user manipulate the widgets which are associated with functions to perform computational tasks based on the state of the widget.

© The Author(s), under exclusive license to Springer Nature Switzerland AG 2023
J. Wang and A. Wang, *Introduction to Computation in Physical Sciences*, Synthesis Lectures
on Computation and Analytics,
https://doi.org/10.1007/978-3-031-17646-3_3

3.1.1   Widgets in Jupyter Notebooks

We discuss `ipywidgets` first because it is easier to build a functional user interface with minimum delay. Follow Program A.1.1 to manipulate the widgets discussed next.

To use `ipywidgets`, we import it first and rename it to `wgt`

```
In [1]: import ipywidgets as wgt
```

Next let us create a button object assigned to the variable `b`

```
In [2]: b = wgt.Button(description='a button')
        b
```

We should see a gray button like the one shown in Fig. 3.1. The button object takes only a description name. A button becomes functional when it is bound to a function to do something useful,

```
In [3]: def hello(dummy):
            print('hello world')

        b.on_click(hello)
```

Here we define a function `hello` that just prints "hello world" when called. We bind it to the button via the on_click() method by passing the name `hello` to it. Now if we click on the button, we should see `hello world` printed below the button. The dummy argument needs to be there because a button object is passed to `hello` which just ignores it.

To choose an item from a menu, we have the select widget,

```
In [4]: menu = wgt.Select(options=['a', 'b', 'c'], description='select menu')
        menu
```

The `Select` menu gets its list of items from `options`, and its name from the description. The selected item can be obtained with the value attribute,

```
In [5]: menu.value
```

```
Out[5]: 'a'
```

All items in the select menu are visible at once as shown in Fig. 3.1. The dropdown menu is functionally the same except it shows only the selected item

Fig. 3.1 Sample Jupyter notebook widgets. From left to right: a button, selection menu, radio buttons, a check box, and a float slider

```
In [6]:  drop = wgt.Dropdown(options=['a', 'b', 'c'], description='dropdown')
         drop
```

Again, the menu item can be read from the value attribute. Yet another alternative method of menu selection is accomplished with the radio button,

```
In [7]:  rb = wgt.RadioButtons(options=['a', 'b', 'c'], description='radio')
         rb
```

The radio buttons are more compact compared to the select menu (see Fig. 3.1). For binary choices, the checkbox widget can be used,

```
In [8]:  chk = wgt.Checkbox(description='yes')
         chk
```

Its value will return either true or false

```
In [9]:  chk.value
```

```
Out[9]:  False
```

The toggle button does the same thing,

```
In [10]:  tb = wgt.ToggleButton(description='boolean')
          tb
```

The button changes its shade between clicks.
For continuous values, the slider widget is useful,

```
In [11]:  slider = wgt.FloatSlider(min=0, max=1, step=0.01, value=0.5)
          slider
```

The input parameters to the `FloatSlider` are self explanatory. The initial value can also be specified. Like menu widgets, its value is read from the value attribute

```
In [12]:  slider.value
```

```
Out[12]:  0.5
```

An equivalent widget exists for integers,

```
In [13]:  intslider = wgt.IntSlider(min=0, max=10, step=1, value=5)
          intslider
```

3.1.2 Interactive Computing

Now we can combine some of the widgets to do something meaningful. Jupyter notebook makes it easy through the `interactive` function which automatically builds appropriate widgets. Here is an example for interactive plotting,

```
In [14]:   import matplotlib.pyplot as plt
           import numpy as np
           def plotf(k, func, refresh):
               if (func == 'sin'):
                   f = np.sin(k*np.pi*x)
               else:
                   f = np.cos(k*np.pi*x)
               if refresh:
                   plt.plot(x, f)

           x = np.linspace(-1,1,101)
           wgt.interactive(plotf, k=(0,2,.1), func=['sin', 'cos'], refresh=(True))
```

In the above example, we define a function `plotf` to graph either $\sin(k\pi x)$ or $\cos(k\pi x)$ depending on the parameter `func`. The `plotf` function is passed to `interactive` along with the parameters. Based on the type of parameter, the `interactive` function will build an appropriate widget to obtain its value, see Fig. 3.2. Here, the first parameter k is inferred to be a real value because it is passed with $k = $ (min, max, step), so a float slider is chosen. The second parameter `func` is a list, so `interactive` draws a select menu of two items. The third parameter `refresh` is a boolean, and a checkbox is used. All this is automatically done for us.

The `interactive` function effectively binds the `plotf` function to the widgets. When any widget value is changed, the user-defined function is called. In this case, a sine or cosine function is plotted based on the values of k and the function selection. The boolean parameter `refresh` controls whether actual graphing is carried out. This can be useful to avoid lagging in complex calculations because when the slider is being dragged to a new value, every intermediate value in-between triggers fresh calculation and plotting. If `refresh` is set to false, we can modify the values of k and other parameters without having to redraw the curve every time.

We can also use a manual switch to `interactive` to achieve a similar goal. Insert `'manual':True` as the second argument to `interactive` after `plotf` in Program A.1.1 as

Fig. 3.2 Interactive plotting with widgets. Either $\sin(k\pi x)$ or $\cos(k\pi x)$ is graphed depending on the parameters k and the type of function. See text

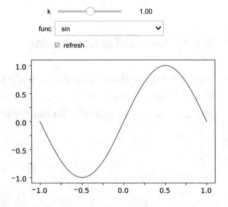

```
In [15]:  intslider
          wgt.interactive(plotf, {'manual':True}, k=(0,2,.1), func=['sin', 'cos'], refresh=(True))
```

We will get another button labeled `Run Interact` underneath the refresh button. As we change the slider value or the function selection, the figure is not drawn unless we click on the run button, even if the refresh button is checked. The latter is redundant in this case, and can be removed from `plotf` and from `interactive`.

3.1.3 Python Widgets

Interactive widgets are also available in standard Python through `matplotlib`. Except for menus, they are similar to the IPywidgets counterparts such as sliders and buttons. Unlike IPywidgets, the `matplotlib` library does not have an interactive function which automatically builds widgets within the context of parameters. Users must place the widgets on the plot area. It takes time to arrange everything in the right place, but once you have a template, the time can be minimized. But the advantage is that the interaction is more responsive and less lagging, and that the user has more control.

Program A.1.2 does functionally the same job as Program A.1.1. To build radio buttons, for instance, this code is used

```
In [16]:  fs  = ['sin', 'cos']          # radio buttons
          rbox =fig.add_axes([.15, .02, .15, .15])  # list = [x, y, dx, dy]
          func = RadioButtons(rbox, fs, 0)   # active = sin (0)
```

Like before, the function choices are contained in the variable `fs`. The variable `rbox` draws a box for the radio buttons at one corner $(x, y) = (0.15, 0.02)$ extended to $dx = dy = 0.15$ at the other corner. Here the coordinates refer to the plot area which has the origin at $(0, 0)$ and extends to $(1, 1)$ in x and y directions. Actual buttons are drawn by the next statement with `RadioButtons`. The layout is very similar to Fig. 8.10. In fact, Program A.1.2 uses Program A.6.6 as a template. As usual, the buttons `func` is bound to a function that is called when it is clicked (see Program A.1.2).

3.2 Interactive Visualization and Animation

The power of visualization cannot be understated in scientific computing. Simulations and data come alive when accompanied with animation because it visually engages the participant and helps with understanding the underlying physical processes, and even the algorithms sometimes. Interactive visualization deepens the engagement and understanding. We briefly introduce two methods of visualization that are used throughout, Matplotlib animation and `vpython`.

3.2.1   Matplotlib Animation

As described earlier (Sect. 2.10), Matplotlib is an easy-to-use yet powerful plotting package in Python. It is equally capable of animation. The basic process is to either change the data of a plot incrementally or generate all data sets at once to be shown sequentially. We use `FuncAnimation` for the former and `ArtistAnimation` for the latter. The two are equivalent, and our programs use both (see Sect. 9.1.1).

To be concrete, we will discuss `FuncAnimation` with an example (Program A.1.3). Suppose we wish to show a harmonic wave in real time, $u(x, t) = \sin(x) \sin(\omega t)$. First we need several libraries, particularly the Matplotlib `animation` module aliased to am,

```
In [17]:  import matplotlib.pyplot as plt, numpy as np
          import matplotlib.animation as am
          %matplotlib notebook
```

The magic function `%matplotlib notebook` is necessary so animation works in Jupyter notebook. It is not needed in a regular Python code. Next we define a function to update the figure data,

```
In [18]:  def updatefig(*args):              # args[0] = frame = t
              t = 0.1*args[0]
              u = np.sin(x) * np.sin(w*t)
              plot.set_data(x, u)             # update data
```

This function will be input to and called by `FuncAnimation` as seen shortly. Upon entering `updatefig`, the first argument `args[0]` will be a counter registering the number of calls made to the function, so we use that as some scaled timer. The data u is calculated, and finally it is set to the plot.

The plot is refreshed with `FuncAnimation`,

```
In [19]:  w = 1.0
          x = np.linspace(0, 2*np.pi, 101)
          fig = plt.figure()
          plot = plt.plot(x, 0*x)[0]              # create plot object
          ani = am.FuncAnimation(fig, updatefig, interval=5, blit=True)    # animate
          plt.ylim(-1.2,1.2);
```

After setting ω and the x grid, we create the figure and initiate the plot object assigned to the variable `plot`.[1] We pass the figure `fig` and most important, the updater `updatefig` to `FuncAnimation`, which periodically refreshed the plot. We should see a pulsating sinusoidal wave. The `blit` parameter uses hardware acceleration, and can be omitted if it causes problems. Animation with `ArtistAnimation` can be done similarly. See Program A.6.2, Program A.7.1 and others in Appendix A for more Matplotlib animations.

---

[1] A Matplotlib object returns a tuple of sub-objects and we need only the first one, hence the `[0]` index.

3.2.2   VPython, 3D Visualization

VPython is a graphics library that makes 3D animation easy.[2] It is a popular tool in physics and mathematics education communities that integrate computation into the curriculum. We use it in the chapters ahead, but let us dive right in with an example Program A.1.4. We discuss running VPython locally next but it can also be run in the cloud with GlowScript (see Sect. 1.3).

We import vpython like any other library,

```
In [20]:  import vpython as vp
```

If it generates an error message, it is likely because vpython is not part of standard Python distributions like Anaconda as of yet, so we need to install it first. Assuming your base installation is Anaconda, open the Anaconda prompt (terminal) and type
conda install -c vpython vpython
If you have a custom Python installation, install vpython with
pip install vpython
Next we create a scene with a couple of 3D objects, a sphere and a box,

```
In [21]:  scene = vp.canvas(title=('right drag to change camera,' +
                            'double drag to zoom'))
          ball = vp.sphere()
          box = vp.box(pos=vp.vector(0,0,-1.1), length=10, height=5, width=.2,
                            color=vp.color.cyan)
```

Graphics output is rendered via a web page, so we should see a window appearing in your default web browser. We used canvas explicitly to create the scene and add a title to the window, but it is not necessary because a scene is created by default.

The first object drawn is a sphere, with default attributes and assigned to the variable ball so it can be manipulated later on. The second object is a box, with some specific attributes to the box object. A vpython object can have a number of attributes, including position, size, color, and so on. Some attributes like position refer to the coordinate system of the scene. By default, we can look at the computer screen as the $x - y$ plane, with the x direction to the right and y direction up, and z direction coming out of the screen. The default position of ball (center of sphere) is at the origin, so the sphere is in front of the thin box serving as the background. There are other objects in vpython one can use, shown in Fig. 3.3.

We can interact with the scene as the window title suggests: swing the camera around with the mouse while holding its right button, or zoom in or out holding both buttons.

To make the objects move, all we have to do is changing the position. Again, let us assume the ball moves as a harmonic oscillator, the following will cause it to move sinusoidally,

---

[2] https://www.vpython.org.

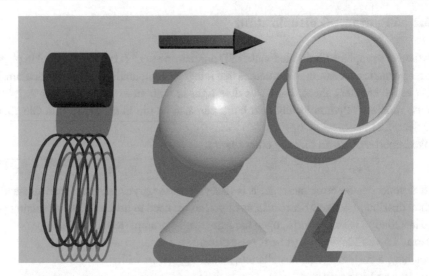

Fig. 3.3 Select vpython objects. Clockwise from upper left, they are created with cylinder, arrow, ring, pyramid, cone, and helix. At the center is the sphere. The objects are post-processed with POV-Ray to include ray tracing effects

```
In [22]:  t = 0
          while t<10:
              vp.rate(10)
              ball.pos = vp.vector(2*vp.sin(2*t),0,0)
              t += 0.1
```

We update the ball.pos attribute as a function of time, and the scene is animated with vp.rate. The latter also controls the speed of animation (10 frames per second here). The camera angle and zoom can also be changed while the animation is being played.

Many of our programs either have vpython integrated into them or can be readily made so. See Program A.3.4 and others in Appendix A for more examples of vpython.

GlowScript and Trinket animation

GlowScript is a language and platform that behaves like VPython but runs directly in a web browser (see Sect. 1.3). Usually we need to make slight modifications of vpython programs to run in GlowScript, such as replacing vector with vec. But the advantage is that we do not need a Python installation to run, and simulations can be shared with a simple web link. See select GlowScript programs available at

- https://www.glowscript.org/#/user/jaywang/folder/intro/

Also see Appendix A for examples.

Trinket is another platform similar to GlowScript, but it is easier to modify and share Trinket programs. One can also embed Trinket program within a web page. A user with a Trinket link can run and modify the program, create a new link to the modified version, and share it with others. See enclosed link in Appendix A for examples.

3.3 Sympy

Scientific computing relies on well-established, common libraries. In Python, these include the best-known ones, Sympy, Numpy, and Scipy. Additionally, Numba can help with optimization to speed up calculation with little effort on our part. Each comes with thorough documentation. We discuss each briefly, beginning with Sympy,

Sympy is a symbolic Python library for algebraic computation with symbols and variables.[3] It helps us focus on the problem-solving part and avoid being side tracked with details such as tedious or laborious derivations or integrals (others might call it math drudgery). It can also help find solutions that may not be evident manually. Compared to packages such as Mathematica or Maple, Sympy is open source and works natively and seamlessly in Python. Below we introduce a subset of sympy features with examples, all of which are contained in Program A.1.5.

3.3.1 Mathematical Modules

Sympy contains mathematical modules and application modules. We begin with the former.

Constants
Before getting started, we import sympy as any other package,

```
In [23]: from sympy import *
         init_printing()   # pretty print math symbols
```

The init_printing enables LATEX formatted math output in Jupyter notebook.

Sympy contains pre-defined mathematical and physical constants. The common constants are pi, E, I, and oo (double 'o'),

```
In [24]: pi, E, I, oo
```

Out[24]: (π, e, i, ∞)

---

[3] https://www.sympy.org. One can run Sympy without installation at https://live.sympy.org.

We should not redefine or override these constants. As seen earlier (Sect. 2.1.2), the imaginary unit *i* is also defined as `1j` in Python. It is cleaner to use `I` in `sympy` so its algebraic properties are retained.

Basic Symbolic Algebra and Calculus

To be able to manipulate symbols and variables, we first need to declare them as the algebraic type, unlike standard Python variables that get their types from assignment,

```
In [25]:  x, a, b, c = symbols('x a b c')
          y = a*x*x + b*x + c
          y
```

Out[25]: $ax^2 + bx + c$

The `symbols` module returns variables `x`, `a`, `b`, `c` that are represented by string symbols *x, a, b, c* in quotes. The string symbols can use Greek symbols such as `'omega'` for *ω* (see below).[4] Next, a quadratic expression is assigned to variable `y` as a standard assignment. Note that we cannot omit the multiplication sign such as `b*x` to produce *bx*, not just b x because spaces are just separators in a programming language.

We can manipulate the expression `y` like any algebraic expression. We can differentiate *dy/dx* with respect to *x*, for instance,

```
In [26]:  dydx = Derivative(y,x)
          dydx
```

Out[26]: $\frac{\partial}{\partial x}(ax^2 + bx + c)$

`Derivative` outputs the symbolic derivative operation, but not the actual derivative. To obtain it explicitly, we use the `doit()` method,

```
In [27]:  dydx.doit(), y.diff(x)
```

Out[27]: $(2ax + b, \quad 2ax + b)$

The `y.diff(x)` means taking derivative of of `y` and gives the same result of course. Still, the `Derivative` function is useful sometimes to show the symbolic operation.

The inverse operation to differentiation is integration, obtained with

```
In [28]:  integrate(y, x)
```

Out[28]: $\frac{ax^3}{3} + \frac{bx^2}{2} + cx$

The same results may be obtained with the `integrate()` method,

```
In [29]:  y.integrate(x)
```

---

[4] One could also import all letter and Greek symbols with `from sympy.abc import *`. The advantage is we have no need to define symbols individually. The disadvantage is that it can override default variables such as mathematical constants stated earlier, unless it is imported before anything else as the very first line. Even then, it is common that a variable has been overwritten as something else via assignment, so we have to declare it again via `symbols` if we wish to use it to represent something else. So by all means use the abc module to save typing, but be careful with overwriting variables. It is the adage again: with power comes responsibility.

as well as with

```
In [30]: Integral(y, x).doit()
```

Like its counterpart, `Integral` denotes symbolic integration but does not actually carry out the operation, so we use the `doit()` method. The integration constants are omitted in these indefinite integrals.

Algebraic Solvers

Many a time we need to solve an equation for the roots. Sympy has several for the task. The simplest is `solve`,[5]

```
In [31]: solve(y,x)
```

$$\text{Out[31]:} \quad \left[\frac{1}{2a}\left(-b+\sqrt{-4ac+b^2}\right), \quad -\frac{1}{2a}\left(b+\sqrt{-4ac+b^2}\right)\right]$$

It takes as input the equation to be solved, and the variable to solve for. The expression is implicitly assumed to be equal to zero, $y(x) = 0$. We have the well-known roots to a quadratic equation.

Another example involves the roots of $x^3 - 1 = 0$,

```
In [32]: solve(x**3-1,x)
```

$$\text{Out[32]:} \quad \left[1, \quad -\frac{1}{2}-\frac{\sqrt{3}i}{2}, \quad -\frac{1}{2}+\frac{\sqrt{3}i}{2}\right]$$

We obtain three roots along the unit circle in the complex plane, $\exp(i2n\pi/3)$, $n = 0, 1, 2$. We can verify one of the roots by an expansion,

```
In [33]: expand(E**(4*pi*I/3), complex=True)
```

$$\text{Out[33]:} \quad -\frac{1}{2}-\frac{\sqrt{3}i}{2}$$

We made use of the predefined constants in `sympy`. The parameter `complex=True` tells `expand` to output the results as real plus imaginary parts.

Differential Equation Solver

Sympy also has a rudimentary differential equation solver. To use it, we need a function variable type,

```
In [34]: f = Function('f')
         k = symbols('k')
         dsolve(diff(f(x),x)-k*f(x),f(x))  # diff(f(x),x)==f(x).diff(x)
```

$$\text{Out[34]:} \quad f(x) = C_1 e^{kx}$$

The `Function` module is similar to `symbols` except it returns a function type that can take symbolic arguments like `f(x)`. Like `solve`, `dsolve` takes two inputs, the differential equation as $df(x)/dx - kf(x) = 0$, and the function $f(x)$ to solve for.

---

[5] Sympy has another powerful solver `solveset`, see Sect. 9.2.

Same results are obtained if `diff` is replaced with `Derivative`,

```
In [35]: dsolve(Derivative(f(x),x)-k*f(x),f(x))
```

A second-order differential equation $d^2f/dx^2 = -k^2f$, can be solved similarly,

```
In [36]: dsolve(diff(f(x),x,x)+k*k*f(x),f(x))
```

Out[36]: $f(x) = C_1 e^{-ikx} + C_2 e^{ikx}$

Note the second derivative is specified by `diff(f(x),x,x)` with two x variables in a row.

Definite Integrals

Calculation of definite integrals is also one of the most common tasks in scientific computation like in quantum mechanics. It is done by specifying limits in indefinite integrals,

```
In [37]: integrate(sin(x)**2/x**2,(x,-oo,oo))
```

Out[37]: π

The variable oo is `sympy`'s way of denoting infinity. Sympy can navigate quasi-singularity in $\int_{-\infty}^{\infty} \sin^2(x)/x^2\,dx$ gracefully. It can also handle simple integrable singularities like $\int dx/\sqrt{x}$ without a hiccup,

```
In [38]: integrate(1/sqrt(x), (x,0,1))
```

Out[38]: 2

However, some apparently simple integrals in `sympy` are anything but simple, such as this integral of a Lorentzian (see Exercise 3.6),

```
In [39]: res = integrate(1/(x*x+a*a)**2,(x,-oo,oo))
         simplify(res)
```

Out[39]: $\begin{cases} \frac{\pi}{2a^3} & \text{for } \left|\text{periodic}_{\text{argument}}\left(\frac{1}{\text{polar\_lift}^2(a)},\infty\right)\right| < \pi \\ \int_{-\infty}^{\infty} \frac{1}{(a^2+x^2)^2}\,dx & \text{otherwise} \end{cases}$

The function `simplify` can be very helpful in reducing the results to simple terms, but it fails in this case (see also Sect. 6.1.2). The difficulty here is that `sympy` is very careful about making assumptions. The constant a in the integrand $1/(x^2 + a^2)^2$ is meant to be real and positive, but `sympy` doesn't assume that. So we need to give it a little help if we want the results to be as simple as possible.

The way to do so is to be explicit in declaring the symbol a, like the following,

```
In [40]: a = symbols('a', positive=True)   # specify conditions to help simplify
         integrate(1/(x*x+a*a)**2,(x,-oo,oo))
```

Out[40]: $\frac{\pi}{2a^3}$

By specifying `positive=True`, it knows that a is real and positive, so `sympy` has no need to make assumptions on the argument of a in case it was complex.

Often, in thermodynamics we need integrals of the form,

```
In [41]: p=0.5
         res = Integral(x**p/(exp(x)-1),(x,0,oo))
         res
```

Out[41]: $\int_0^\infty \frac{x^{0.5}}{e^x-1}\,dx$

These integrals are needed in blackbody radiation or in Bose-Einstein condensation. They are not analytically reducible to more convenient terms, so a numerical value is often substituted,

```
In [42]: N(res), res.n(10)
```

Out[42]: (2.31515737339412, 2.315157373)

The N function and the n() methods give the same result. The latter has the advantage of being able to specify the number of significant digits, 10 in this case.

Gaussian integrals are also common, and can be obtained with,

```
In [43]: n = symbols('n', integer=True, positive=True)
         res = integrate(exp(-x*x) *x**n, (x,0,oo))
         res
```

Out[43]: $\frac{1}{2}\Gamma(\frac{n}{2}+\frac{1}{2})$

The result is given in terms of the special Γ function (see later). To obtain results specific to n, we use the substitute method subs. The following gives the value for $n = 2$,

```
In [44]: res.subs(n,2), N(res.subs(n,2))
```

Out[44]: $\left(\frac{\sqrt{\pi}}{4}, \quad 0.443113462726379\right)$

Note the non-numerical and numerical formats.

Expansion Coefficients

Calculations of expansion coefficients is often necessary in quantum mechanics. Sympy can simplify these tasks considerably.

Suppose we wish to expand the initial wave function $\psi(x) = x(a - x)$ in the infinite potential well (see Sect. 7.3.1) as

$$\psi(x) = \sum_n c_n \psi_n(x), \quad c_n = \int_0^a \psi(x)\psi_n^*(x)\,dx. \tag{3.1}$$

In sympy, the result is,

```
In [45]: psi = lambda x: x*(a-x)
         psin = lambda n, x: sqrt(2/a)*sin(n*pi*x/a)
         a = symbols('a', positive=True)
         A = Integral(psi(x)**2,(x,0,a))   # norm
         cn = Integral(psi(x)*psin(n, x),(x,0,a))/sqrt(A)
         cn
```

Out[45]: $\frac{1}{\sqrt{\int_0^a x^2(a-x)^2\,dx}}\int_0^a \frac{\sqrt{2}\sqrt{x}}{\sqrt{a}}(a-x)\sin\left(\frac{\pi x}{a}n\right)dx$

```
In [46]: cn = cn.doit()
         cn
```

Out[46]: $\frac{\sqrt{30}}{a^{\frac{5}{2}}} \left(-\frac{2(-1)^n \sqrt{2} a^{\frac{5}{2}}}{\pi^3 n^3} + \frac{2\sqrt{2} a^{\frac{5}{2}}}{\pi^3 n^3} \right)$

The normalization constant A needs to be calculated because the wave function $\psi(x)$ is unnormalized. After simplification, we have

```
In [47]: simplify(cn)
```

Out[47]: $\frac{4\sqrt{15}}{\pi^3 n^3} \left(-(-1)^n + 1 \right)$

Note that dimensioned parameters such as a drop out because the coefficients c_n are dimensionless. It shows that the results are nonzero only if n is odd. This is due to the symmetry of the initial wave function. See Exercise 3.5 for a nonsymmetric wave function.

One property of the coefficients is that c_n^2 represents the probability of finding the state $\psi_n(x)$, so all probabilities should add up to one, $\sum_n c_n^2 = 1$. Let us go ahead and check,

```
In [48]: res = Sum(cn**2, (n,1,oo))
         res
```

Out[48]: $\sum_{n=1}^{\infty} \frac{240}{\pi^6 n^6} \left(-(-1)^n + 1 \right)^2$

It is just a summation, unevaluated. To evaluate it, we use the `doit` method,

```
In [49]: res.doit()
```

Out[49]: $-\frac{480}{\pi^6} \mathrm{Li}_6 \left(e^{i\pi} \right) + \frac{32}{63}$

The result is expressed in terms of a special function called logarithmic integral that we do not want to get into here.[6] Using `simplify` on the result also does not do much, so let us just evaluate it numerically,

```
In [50]: res.doit().n()
```

Out[50]: 1.0

This is the expected result.

Special Functions

Historically, special functions (see Boas 2006) had been developed analytically for the main part, but they continue to play an important role even in the age of digital computing. We have already seen the Γ function earlier. Sympy has an extensive library of special functions. Here are a few examples,

```
In [51]: erf(x), gamma(x), airyai(x), jn(n,x)
```

Out[51]: $(\mathrm{erf}(x), \quad \Gamma(x), \quad Ai(x), \quad j_n(x))$

---

[6] Those who still keep tables of sums and integrals around (see Abramowitz and Stegun 1970, Gradshteyn and Ryzhik 1980) may recall that $\sum_{n=1,3,5\ldots} 1/n^2 = \pi^2/8$, $\sum_{n=1,3,5\ldots} 1/n^4 = \pi^4/96$, $\sum_{n=1,3,5\ldots} 1/n^6 = \pi^6/960$. The last sum is directly applicable to our case.

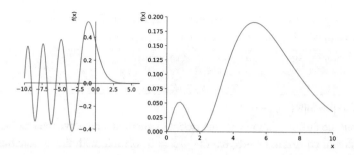

Fig. 3.4 Output of sympy's plot function. Left: the Airy function $Ai(x)$. Right: Radial probability distribution of hydrogen atom $R_{20}^2(r)r^2$

They are respectively, the error $\mathrm{erf}(x)$, gamma $\Gamma(x)$, Airy $Ai(x)$, and spherical Bessel $j_n(x)$ functions.

We will encounter a few next. To get a feel of what they look like, we can plot them with plot,

```
In [52]: plot(airyai(x)) # may need to run twice
```

The $Ai(x)$ is shown in Fig. 3.4 (left). By default the plot uses adaptive sampling to output a smooth curve. The plot function is handy to graph any algebraic expression.

Some special functions are expressed in integral form,

```
In [53]: integrate(exp(-x*x),(x,0,a))  # error func
```

Out[53]: $\frac{\sqrt{\pi}}{2}\,\mathrm{erf}\,(a)$

Sympy even has a Dirac function $\delta(x - a)$,

```
In [54]: integrate(f(x)*DiracDelta(x-a),(x,-oo,oo))
```

Out[54]: $f(a)$

Another special function that had been forgotten for a while but has found resurgence recently in many applications is the Lambert W function, which is defined as the solution to the equation $We^W - x = 0$,

```
In [55]: W = symbols('W')
         solve(W*exp(W)-x, W)
```

Out[55]: [LambertW (x)]

Sympy names this function LambertW.

As an example, Wien's displacement law in blackbody radiation can be solved in terms of the W function. The solution was thought to exist in numerical form only. Just recently it was found to be solvable using the W function. The blackbody radiation spectrum in frequency ν is given by $f(x) = x^3/(\exp(x) - 1)$, $x = h\nu/kT$. The peak frequency occurs at $df/dx_m = 0$, which gives the relation $\nu = CT$, $C = kx_m/h$.

```
In [56]:   eqn = Derivative(x**3/(exp(x)-1),x)
           res = solve(eqn.doit(),x)   # freq, x = hv/kT
           res
```

Out[56]: $\left[\text{LambertW}\left(-\frac{3}{e^3}\right) + 3\right]$

We carry out the differentiation with the `doit()` method, and solve the resultant, which is a transcendental equation, $(x - 3)\exp(x) + 3 = 0$ solvable with the W function. Note the solutions are given as a list of one element.

Numerically,

```
In [57]:   N(res[0])
```

Out[57]:   2.82143937212208

We use `res[0]` because it is in a list even though there is only one solution. So we have Wien's law in frequency form $\nu = 5.88 \times 10^{10}T$ Hz/K. One can obtain in a similar way its wavelength form (see Exercise 3.4).

Linear Equations and Eigenvalues

Most matrix and linear systems of equations are handled numerically. But it is still instructive, even desirable, to manipulate such problems algebraically with low dimensional systems to understand how certain variables scale (see Sect. 6.1) or how the processes unfold.

Let us look at a two-equation problem. Now we introduce some variables with subscripts,

```
In [58]:   x, y = symbols('x y')
           a1, a2 = symbols('a1 a2')
           b1, b2 = symbols('b1 b2')
           c1, c2 = symbols('c1 c2')
           eqns = [a1*x + a2*y-c1,
                   b1*x + b2*y-c2]
           eqns
```

Out[58]: $[a_1 x + a_2 y - c_1, \quad b_1 x + b_2 y - c_2]$

The number following a variable string is printed as a subscript. The expression `eqns` contains two linear equations with unknowns x and y. They are solved as

```
In [59]:   res = solve(eqns, x, y)
           res
```

Out[59]: $\left\{x : \frac{-a_2 c_1 + b_2 c_1}{a_1 b_2 - a_2 b_1}, \quad y : \frac{a_1 c_2 - b_1 c_1}{a_1 b_2 - a_2 b_1}\right\}$

We specify two unknowns to `solve` instead of just one.

A more familiar way is to use matrix notation

```
In [60]: A = Matrix([[a1,a2],[b1,b2]])
         A
```

Out[60]: $\begin{bmatrix} a_1 & a_2 \\ b_1 & b_2 \end{bmatrix}$

`Matrix` takes rows (list of lists) as input and stack them up to form a matrix. If a single list is given, it is construed as a column matrix. To solve a linear system of the form $A\mathbf{x} = \mathbf{c}$ where $\mathbf{x}$ and $\mathbf{c}$ are column matrices, we have the following,

```
In [61]: solve(A*Matrix([x, y])-Matrix([c1, c2]), Matrix([x, y]))
```

Out[61]: $\left\{ x: \frac{-a_2c_2+b_2c_1}{a_1b_2-a_2b_1}, \quad y: \frac{a_1c_2-b_1c_1}{a_1b_2-a_2b_1} \right\}$

To demonstrate eigenvalue problems, we start with a symmetric matrix (Hermitian),

```
In [62]: B = A.subs([[b1,a2], [b2,a1]])    # symmetric matrix
         B
```

Out[62]: $\begin{bmatrix} a_1 & a_2 \\ a_2 & a_1 \end{bmatrix}$

where the substitutions are contained with a pair of lists so $b_1 \to a_2$ and $b_2 \to a_1$.

The eigenvalues and eigenvectors are obtained with the `eigenvects` method,

```
In [63]: B.eigenvects()
```

Out[63]: $\left[\left(a_1 - a_2, \quad 1, \quad \left[\begin{bmatrix} -1 \\ 1 \end{bmatrix} \right] \right), \quad \left(a_1 + a_2, \quad 1, \quad \left[\begin{bmatrix} 1 \\ 1 \end{bmatrix} \right] \right) \right]$

The results are given as a list of tuples, each containing the eigenvalue, degeneracy (multiplicity), and unnormalized eigenvector.

3.3.2   Application Modules: Physics

Sympy has also several modules for specific physical systems. We now discuss select modules designed to explore the systems symbolically.

Quantum Simple Harmonic Oscillator

The quantum simple harmonic oscillator is the most important and elegant problem in quantum mechanics. It can be explored in `sympy` meaningfully without getting too much into the mathematical details.

We start with importing specific modules and some definitions,

```
In [64]: from sympy.physics.qho_1d import E_n, psi_n, hbar
         n = symbols('n', integer=True, positive=True)
         m, omega = symbols('m omega', positive=True)
         x, a = symbols('x a')
         E_n(n, omega), psi_n(0,x,m,omega)
```

Out[64]: $\left(\hbar\omega\left(n+\frac{1}{2}\right), \quad \frac{\sqrt[4]{m}\sqrt[4]{\omega}}{\sqrt[4]{\hbar}\sqrt[4]{\pi}}e^{-\frac{m\omega x^2}{2\hbar}}\right)$

In addition to the energy E_n and wave function ψ_n, we also imported the constant $\hbar$ (Eq. (7.6))
from the qho_1d module. It also shows the ground state wave function explicitly.

Unlike the classical oscillator where the particle's motion is restricted to the region
between turning points (amplitudes) for a given energy, the quantum oscillator has a nonzero
probability even in the classically forbidden region. We can explore this with the module to
find the probability of the particle in the classically forbidden region. The turning points are
defined as when the energy is equal to the potential energy $V(x) = E$. In sympy we solve
for them as

```
In [65]: V = m*omega**2 * x**2/2
         tp = solve(E_n(0, omega)-V, x)   # turning points
         tp
```

Out[65]: $\left[-\frac{\sqrt{\hbar}}{\sqrt{m}\sqrt{\omega}}, \quad \frac{\sqrt{\hbar}}{\sqrt{m}\sqrt{\omega}}\right]$

The two points are symmetric about the origin as expected. The probability to one side of
the forbidden region is found with

```
In [66]: prob = Integral(psi_n(0,x,m,omega)**2, (x,tp[1],oo))
         res = prob.doit()
         res
```

Out[66]: $\frac{1}{2\sqrt{\pi}}\left(-\sqrt{\pi}\,\mathrm{erf}\,(1)+\sqrt{\pi}\right)$

The result is in terms of the error function seen earlier. Doubling the result to account for
the other side, it is numerically

```
In [67]: 2*N(res)
```

Out[67]: 0.157299207050285

It is about 16% in the ground state. It is expected to decrease with higher excited states due
to the semiclassical limit (Exercise 3.7).

Hydrogen Atom

The hydrogen atom is the first success of quantum mechanics in explaining the experimentally observed properties of an actual physical system, and it validated quantum mechanics
as a fundamentally correct theory.

The energy and wave functions of the hydrogen atom are found in the hydrogen
module,

```
In [68]: from sympy.physics.hydrogen import E_nl, R_nl
         n, l = symbols('n l', integer=True, positive=True)
         Z, r = symbols('Z r', positive=True)
         E_nl(n,Z)
```

Out[68]: $-\frac{Z^2}{2n^2}$

The energy E_n is given in atomic units (see Sect. 7.3.3), and depends on the quantum number n only, not on l as an exception rather than the rule. The wave function depends on both nl, for instance,

```
In [69]: R_nl(2,0,r)
```

Out[69]: $\frac{\sqrt{2}}{4}(-r+2)e^{-\frac{r}{2}}$

The radius r is also given in atomic units for length (Bohr radius). We can graph the radial probability density with

```
In [70]: plot((r*R_nl(2,0,r))**2, (r,0,10))   # may have to run the cell again
```

The output is shown in Fig. 3.4.

We can also calculate expectation values of the radius $\langle r \rangle = \int R_{nl}^2(r)\, r\, r^2 dr$ in a given state,

```
In [71]: res = Integral(R_nl(2,0,r)**2*r**3,(r,0,oo))  # <r>
         res
```

Out[71]: $\int_0^\infty \frac{r^3}{8}(-r+2)^2 e^{-r}\, dr$

```
In [72]: res.doit()
```

Out[72]: 6

The expectation value of the inverse radius $\langle 1/r \rangle$ is

```
In [73]: res = Integral(R_nl(2,0,r)**2*r,(r,0,oo))   # <1/r>
         res.doit()
```

Out[73]: $\frac{1}{4}$

The state $n, l = n, n-1$ is a special state in that the probability is concentrated in a thin spherical shell, called circular states. The expectations of $\langle r \rangle$ and $\langle 1/r \rangle$ are

```
In [74]: integrate(R_nl(n,n-1,r)**2*r**3, (r,0,oo)), integrate(R_nl(n,n-1,r)**2*r, (r,0,oo))
```

Out[74]: $\left(\frac{2\cdot 2^{2n-2}}{n^3(2n-1)!} n^{-2n+2} \left(\frac{2}{n}\right)^{-2n-1} \Gamma(2n+2), \quad \frac{2^{2n-2}n^{-2n+2}}{n^4(2n-1)!} \left(\frac{2}{n}\right)^{-2n+1} \Gamma(2n+1) \right)$

Simplifying the results yields

```
In [75]: simplify(_)
```

Out[75]: $\left(n^2 + \frac{n}{2}, \quad \frac{1}{n^2} \right)$

The underscore argument to `simplify(_)` tells it to simplify the last result. For large n, the value $\langle r \rangle \sim n^2$, signifying that the size of circular states scales as n^2. In fact, it is a good estimate of the size of an atom in state n. On the other hand, the value $\langle 1/r \rangle = 1/n^2$ exactly, regardless of the quantum number l.

3.4   Numpy Library

The numpy library can be said to be the backbone of numerical computation in Python.[7] It provides advanced array capabilities and with the speed of a compiled code. It is used throughout the later chapters. We describe several common features in this section with working template in Program A.1.6.

Constants

The numpy package is conventionally named np,

```
In [76]: import numpy as np
```

It also defines basic mathematical constants often needed, π, e, ∞, and nan,

```
In [77]: np.pi, np.e, np.inf, np.nan
```

```
Out[77]: (3.141592653589793, 2.718281828459045, inf, nan)
```

The type nan means "not a number", and is usually a result of ill-defined operations such as dividing by zero or overflowing. When it appears unintended after a calculation, it means something has very probably gone wrong, and the cause should be tracked down and fixed.

Array Creation

```
In [78]: np.zeros(3), np.ones(3)
```

```
Out[78]: (array([0., 0., 0.]), array([1., 1., 1.]))
```

The modules zeros and ones return arrays initialized to zeros and ones, respectively. The data type is defaulted to double precision floating point numbers, or just floats, though numpy arrays can have other common data types like integer, boolean, text, and so on (Chap. 2). In fact, arange() creates an integer array by default,

```
In [79]: np.arange(3)
```

```
Out[79]: array([0, 1, 2])
```

It is similar to the list with range except it behaves like a numpy array.

Another useful array is that created with linspace,

```
In [80]: np.linspace(-1,1,5,retstep=True)
```

```
Out[80]: (array([-1. , -0.5, 0. , 0.5, 1. ]), 0.5)
```

This example creates an array of five equally-spaced points between $[-1, 1]$ inclusive. Optionally it can return the spacing with retstep=True.

---

[7] https://numpy.org.

We can also create a `numpy` array from existing array-like data,

```
In [81]: a = np.array([1., 2., 3.])
         a
```

```
Out[81]: array([1., 2., 3.])
```

A list is passed to `array`, and `a` now is a three-element array instead of a list.

Array Indexing and Slicing

Like lists (Chap. 2), `numpy` arrays can be accessed with integer indices,

```
In [82]: a[0], a[1], a[-1], a[-2]
```

```
Out[82]: (1.0, 2.0, 3.0, 2.0)
```

We should always be clear that array indexing starts from 0, a negative index counts backward, and the last element is `a[-1]`.

Slicing of arrays can be used to select a range of elements (see Sect. 2.4.2, also Sect. 4.5 for advanced indexing),

```
In [83]: a[1:2]
```

```
Out[83]: array([2.])
```

The `a[m:n]` notation is inclusive of m but exclusive of n, so we get only an array of one element above. If the first index is omitted, the selection starts from the beginning of the array. If the second index is left out, it defaults to the end of the array.

```
In [84]: a[:2], a[2:]
```

```
Out[84]: (array([1., 2.]), array([3.]))
```

Element-Wise Arithmetic

One very useful capability of `numpy` arrays is the element-wise operation.

```
In [85]: b = np.array([2,2,2])
         a + b, a-b
```

```
Out[85]: (array([3., 4., 5.]), array([-1.,  0.,  1.]))
```

Here, an integer array b is added to or subtracted from a, and the operation is done element by element. Such operations are called element-wise, or vector operations. Note that b is an integer array. In numerics, floats are at a higher level than integers as shown earlier (Sect. 2.1.2), so integers are converted to floats in these operations, and the results are in floats.

The same vector operations apply to multiplication and division,

```
In [86]: a * b, a/b
```

```
Out[86]: (array([2., 4., 6.]), array([0.5, 1. , 1.5]))
```

Vector operations circumvent explicit looping, and can be done faster than explicit looping, so it is preferred.[8]

Array Broadcasting

We can even carry out vector operations between a number (scalar) and an array (vector). Consider the following code,

```
In [87]: 2 + a, np.array([2,2,2]) + a
```

```
Out[87]: (array([3., 4., 5.]), array([3., 4., 5.]))
```

In the first instance, we are adding an integer number to an array, and in the second instance, we are adding an integer array. The results are the same. What is happening in the first instance is that the scalar number is effectively (not actually) converted into an integer array of same length, and added to the floats array via vector operation. This conversion is known as broadcasting. Conceptually the scalar is treated as a vector with the same component in each dimension.

Broadcasting does not extend to vectors of different length. The following will generate an error,

```
In [88]: np.array([2,2]) + a
```

```
---------------------------------------------------------------------------
ValueError                                Traceback (most recent call last)
<ipython-input-28-39f07ddf6363> in <module>()
----> 1 np.array([2,2]) + a

ValueError: operands could not be broadcast together with shapes (2,) (3,)
```

Broadcasting with multiplication and division works the same way,

```
In [89]: 2*a, np.array([2,2,2])*a
```

```
Out[89]: (array([2., 4., 6.]), array([2., 4., 6.]))
```

Array Operations

Numpy provides several convenient modules for array operations. The following code sums up the product of two arrays,

```
In [90]: a = np.arange(3)
         b = np.array([2,2,2])
         np.sum(a * b)
```

```
Out[90]: 6
```

---

[8] Occasionally one hears phrases such as the "code is Pythonic." It is subjective and elusive to quantify what is Pythonic, but using vector operations where possible is considered Pythonic. If we were to define Pythonic, we would say a code is Pythonic if it is RARE: Readable And Reasonably Efficient. As long as it gets the job done, style points are secondary.

The result is like the scalar product between two vectors. Indeed, there is a function for that purpose,

```
In [91]: np.dot(a, b)
```

```
Out[91]: 6
```

The difference between the two functions shows when multidimensional arrays are involved. The `dot` behaves like matrix multiplications.

We can take a vector product if the arrays represent the three components of a vector,

```
In [92]: np.cross(a, b)
```

```
Out[92]: array([-2,  4, -2])
```

We in fact use this function in calculating the forces in cross electromagnetic fields (see Program A.3.5).

Sometimes we need to reverse the order of array elements. We can do so with

```
In [93]: np.flip([1,2,3])
```

```
Out[93]: array([3, 2, 1])
```

To insert an element into an array, we use the `insert` function,

```
In [94]: np.insert(a, 1, -1)
```

```
Out[94]: array([ 0, -1,  1,  2])
```

A call to `insert(a, index, value)` inserts a `value` before `index` of the array a. This works the same way with lists (Sect. 2.7.3). The original array is not modified, only a copy is returned. Multiple insertions are also possible. For instance,

```
In [95]: np.insert(a, [1,3], [-1,-2])
```

```
Out[95]: array([ 0, -1,  1,  2, -2])
```

In the above example, the first list contains the indices [1,3] before which the values in the second list [−1, −2] are to be inserted. Because the original array a has only three elements, the second index 3 is interpreted to be after last position for the insertion. So it is inserted at the end (effectively appended).

2D Arrays

Array creation routines described above can be used to create n dimensional arrays.. For instance, to initialize a 3×3 2D array,

```
In [96]: np.zeros((3,3))
```

```
Out[96]: array([[0., 0., 0.],
                [0., 0., 0.],
                [0., 0., 0.]])
```

For matrix calculations, special two-dimensional arrays are required. The following creates a two dimensional array that is diagonal,

```
In [97]: A = np.diag([1., 2., 3.], k=0)
         A
```

```
Out[97]: array([[1., 0., 0.],
                [0., 2., 0.],
                [0., 0., 3.]])
```

The parameter k indicates off-diagonal distance. It can be omitted if it is zero.

For nonzero k, an off-diagonal matrix will be created,

```
In [98]: B = np.diag([1., -1.], k=1)
         B
```

```
Out[98]: array([[ 0.,  1.,  0.],
                [ 0.,  0., -1.],
                [ 0.,  0.,  0.]])
```

The distance of 1 refers to the off-diagonal just above the main diagonal, and it requires two values given in the list. Similarly, an off-diagonal just below the main diagonal can be created with

```
In [99]: C = np.diag([1., -1.], k=-1)
         C
```

```
Out[99]: array([[ 0.,  0.,  0.],
                [ 1.,  0.,  0.],
                [ 0., -1.,  0.]])
```

Incidentally, we can use transpose to obtain the same result,

```
In [100]: np.transpose(B)
```

Combining the three arrays gives us a so-called tridiagonal array,

```
In [101]: A+B+C
```

```
Out[101]: array([[ 1.,  1.,  0.],
                 [ 1.,  2., -1.],
                 [ 0., -1.,  3.]])
```

Element-wise operations are apparent also with 2D arrays.

Functions

Numpy provides most elementary functions. The functions support scalar inputs and vector operations on array inputs. For example, the following gives absolute values,

```
In [102]: x = np.linspace(-1,1,5)
          np.abs(x)
```

```
Out[102]: array([1. , 0.5, 0. , 0.5, 1. ])
```

The function operates on a scalar value as well as element-wise on an array. Such functions are called u-functions, or universal functions.

User-defined function can also be u-functions. We define a quadratic function as follows,

```
In [103]:  def f(x):
               return x*x

           f(x)
```

Out[103]: array([1. , 0.25, 0. , 0.25, 1.])

The function works as expected on an array and on a single value. This is because multiplication of two arrays is a vector operation as seen earlier.

However, if there are operations that are not naturally done element-wise, passing an array to the function can fail. Below is a step function that returns either 1 or 0 depending on the sign of the input. It works with scalar input, but fails with array input because of the `if` statement. To fix it, we use the `vectorize` module,

```
In [104]:  def g(x):
               if x>0:  return 1
               else:    return 0

           gvec = np.vectorize(g)
           gvec(x)
```

Out[104]: array([0, 0, 0, 1, 1])

The vectorized function `gvec` now works on both scalar and vector inputs.

Finally, as described in Chaps. 2 and 8, `numpy` has random number generators that can return either a single value or an array of random values between [0, 1).

```
In [105]:  np.random.random(), np.random.random(3)
```

Out[105]: (0.8887652855521309, array([0.98739968, 0.20369024, 0.84627868]))

We can also generate random integers with

```
In [106]:  np.random.randint(low=0, high=10), np.random.randint(0,10,3)
```

Out[106]: (7, array([0, 5, 1]))

We will use arrays of random numbers in the next section.

3.5 Scipy Library

Scipy is to numerical computation what Sympy is to symbolic computation.[9] These two libraries plus Numpy are complementary and powerful toolboxes for scientific computing in Python. We introduce select functionalities of `scipy` from Program A.1.7 next.

Definite Integrals

As to be discussed later (Sect. 4.2), numerical integration is a common task in scientific computing. Whereas `sympy` can perform analytic integration (Sect. 3.3), many integrals

---

[9] https://scipy.org.

found in practice are too complex to have analytic solutions, and numerical integration is the only way.

Scipy has a robust routine for numerical integration, quad,

```
In [107]:   import numpy as np
            from scipy.integrate import quad
```

It requires us to supply the integrand as a function, and the limits of integration,

```
In [108]:   f = lambda x: 1/np.sqrt(x)
            quad(f, 0, 1)
```

Out[108]:   (2.0000000000000004, 5.10702591327572e-15)

In this example, the integrand $1/\sqrt{x}$ is defined in the lambda function f() (Sect. 2.5), and the limits of integration are [0, 1]. The results are returned as a pair of values: the integral and the estimated absolute error. Here, the result 2 is practically exact, and the error is at the limit of double precision. Note that $x = 0$ is a singularity in the integrand, but it is an integrable singularity. Apparently quad navigates around the this integrable singularity very well.

Scipy can also handle integrals with an open range. The following Lorentzian integral is over the range $[0, \infty)$,

```
In [109]:   a = 1.0
            f = lambda x: 1/(x*x+a*a)**2
            quad(f, 0, np.inf)
```

Out[109]:   (0.7853981633974483, 1.9448610022155054e-09)

Recall from Sect. 3.3, the exact result should be $\pi/4$. The actual result agrees with it to 16 digits, so the estimated error is quite conservative in this case. This is because the integrand decays rapidly with larger x so the integral converges fast.

We are not always that lucky with fast converging integrals. For instance,

```
In [110]:   f = lambda x: np.sin(x)**2/x**2
            quad(f, 0, np.inf)
```

Out[110]:   (1.5708678849453774, 0.0015587759422626135)

This integral generated a warning message from scipy (not shown) about slow convergence. Indeed, compared to the exact result of $\pi/2$ (Sect. 3.3), we now have only four digits of accuracy. This is because this integral converges slowly like $1/a$ for large a, much slower than the Lorentzian integral. As a result, quad has trouble with accuracy. Manual intervention is necessary if higher accuracy is to be gained (see Exercise 3.3).

Expansion Coefficients Revisited

Earlier we obtained the expansion coefficients Eq. (3.1) with sympy for a particle in the infinite potential well. We can almost just copy and paste the same code to do the same calculation numerically in simpy as follows.

```
In [111]: psi = lambda x: x*(a-x)
          psi2 = lambda x: psi(x)**2
          psin = lambda n, x: np.sqrt(2/a)*np.sin(n*pi*x/a)
          f = lambda x: psi(x)*psin(n,x)
          pi = np.pi
          a = 1
          A = quad(psi2, 0, a)[0]    # norm
          for n in range(1,5):
              print(n, quad(f, 0, a)[0]/np.sqrt(A))
```

```
Out[111]: 1 0.9992772459953339
          2 5.870202015831392e-17
          3 0.03701026837019757
          4 -6.221452546292864e-17
```

The only change is replacing the sine and square root functions with Numpy versions[10] and combining the product of two functions `psi` and `psin` into a single `lambda` function because, unlike symbolic computation, the `quad` routine does not parse algebraic expressions and requires a standard function as input. The results show that the coefficient $c_2 = c_4 \simeq 0$ (to double precision) as expected, as well as for all even n. The first term c_1 is dominant, even more so in terms of probability because $c_1^2 = 0.99855$, and the next term is only $c_3^2 = 0.00136$. With these numbers, we could then turn our attention to understanding why (Exercise 3.5) without being too occupied with manual calculations. We leave it to the interested reader to show that the sum of probabilities approaches 1, a result proven algebraically with Sympy (Sect. 3.3).

Differential Equation Solver

Solving ordinary differential equations (ODE) numerically is a very important task in scientific computing. In fact, we will discuss solutions of physical problems expressed in terms of ODEs in Chap. 5. We can also solve ODEs with `scipy` integrators. The general purpose integrator is called `odeint`. To demonstrate the process, suppose we wish to find the velocity as a function of time of a particle under viscous force Eq. (8.2) (or linear drag Eq. (5.12)), $dv/dt = -bv$. Due to its simplicity and because it has an analytic solution Eq. (8.3), it is instructive to solve it numerically and compare with the exact solution.

The `odeint` module requires information about the ODE, which is supplied via a user-defined function. The function will be called by `odeint` to perform the integration. It looks like the following in our example,

```
In [112]: from scipy.integrate import odeint
          def drag(v, t):
              b = 1.0
              return -b*v
```

The `drag` function just returns the right-hand side of the ODE, namely the derivative dv/dt, which depends only on v itself in this case. It could also depend on time in general,

---

[10] Actually we could have used Sympy functions as well, but it is faster to use fully numerical versions of Numpy.

though not in this simple example. Because of this possibility, when odeint calls the function, it will pass the time t as the second argument.

Next, we supply the name of the function to odeint, the initial conditions, and a list of points in time where the solutions are desired,

```
In [113]: t = np.linspace(0,1,5)   # time grid
          v = 1.0                   # initial velocity
          odeint(drag, v, t)
```

```
Out[113]: array([[1.         ],
                 [0.7788008 ],
                 [0.60653067],
                 [0.47236657],
                 [0.36787947]])
```

Internally, the odeint module calls the function drag many times and uses the information returned by it to crank out solutions at the specified instants in time (see Chap. 5). Checking against the exact velocity, the numerical solution is very accurate for this simple problem. To solve a different problem, we only need to change the function itself to return the correct derivative (the right-hand side).

As seen from Eq. (8.2), $dx/dt = v$ and $dv/dt = -bv$ are two coupled ODEs, that is to say, the position x depends on v. We now have the velocity. Can we obtain the position simultaneously? In fact, the ODE solver odeint works perfectly fine for coupled ODEs. What is needed is to supply all the terms on the right-hand side via the user-defined function. For these two ODEs, it is given as

```
In [114]: def drag2(y, t):
              b = 1.0
              v = y[1]
              return [v, -b*v]
          y = [0., 1.]              # y=[x, v], initial pos and velocity
          odeint(drag2, y, t)
```

```
Out[114]: array([[0.        , 1.        ],
                 [0.22119922, 0.77880078],
                 [0.39346933, 0.60653067],
                 [0.52763344, 0.47236656],
                 [0.63212055, 0.36787945]])
```

The new function drag2 looks like the old one, but the first argument y as passed from odeint is a list of position and velocity values $[x, v]$, rather than v only. This is because the initial conditions must have two values in position and velocity, so odeint infers it is a system of two ODEs. It also expects two derivatives from drag2, which correctly returns them in the list $[v, -bv]$. Now the output array has two columns, the first one contains the position, and the second one the velocity, which is the same as the single ODE, of course. The number of rows is equal to the number of points in time.

Special Functions

The scipy library has a full range of special functions as sympy does. It has the usual ones, including

```
In [115]: from scipy.special import erf, gamma, lambertw, spherical_jn
          x = 0.5
          erf(x), gamma(x), lambertw(x)
```

Out[115]: (0.5204998778130465, 1.7724538509055159, (0.35173371124919584+0j))

Note the Lambert *W* function is named `lambertw`, without the capitalization used in `sympy`. As useful and essential as the special functions are, we will briefly encounter two of them in the context of our discussion next.

Equation Solver

There are several root finders in `scipy` including `fsolve` and `bisect`. Consider Wien's displacement law seen earlier (Sect. 3.3). Numerically, the solution to the defining equation is found below,

```
In [116]: from scipy.optimize import fsolve, bisect
          f = lambda x: (x-3)*np.exp(x)+3    # freq, x = hν/kT
          fsolve(f, [1., 3.])
```

Out[116]: array([2.27881059e-12, 2.82143937e+00])

The equation to be solved is given by the `lambda` function, and input as the first argument to `fsolve`. The second argument is a list of initial guesses for roots. They don't have to be precise, but should be as close to the real roots as possible. Equal number of roots are returned by `fsolve`. Of the two guesses, one converged to the trivial root zero, and the other one to the actual root. To have more control over the outcome, the bisection method is better, discussed shortly.

As noted in Sect. 3.3, before the rediscovery of the Lamber *W* function, a numerical solution above was thought to be the only way to solve the problem. With the solution expressed in terms of *W*, we can evaluate it analytically,

```
In [117]: lambertw(-3/np.e**3)+3    # analytic solution
```

Out[117]: (2.8214393721220787+0j)

The zero imaginary part comes about because the *W* function has two real branches and many complex ones, its value is complex in general.

For some equations, the initial guess could lead to a runaway root if it exists, or no root is found at all. A sure way to locate a root that we know to exist in some range is to use the bisection method (Sect. 4.4). In `scipy`, it is invoked with

```
In [118]: bisect(f, 1., 5.)    # [1]=bracket
```

Out[118]: 2.821439372122768

The `bisect` routine uses the same user-defined function. We set the bracket at [1, 5]. In this particular case, we know based on physics that there must be a root between $[\epsilon, \infty)$ with ϵ being an infinitesimal positive number. If we set a bracket without a root inside, the `bisect` will tell us so and quit.

The robustness of the bisection method is very useful in other situations. For example, to determine the zeros of the spherical Bessel functions,[11] the bisection method should be used because of possible runaway roots,

```
In [119]:   f = lambda x: spherical_jn(n,x)
            n = 1
            bisect(f, 1., 5.), bisect(f, 5., 10.)
```

Out[119]:   (4.493409457909365, 7.725251836938014)

Here we are seeing the first two nonzero roots of $j_1(x)$. Bisection is useful because they interleave with those of $j_0(x)$ occurring at multiples of π.

Linear Equations and Eigenvalues

Scipy has complete linear system and eigenvalue solvers. To solve a system of equations $\mathbf{Ax} = \mathbf{c}$, we use solve,

```
In [120]:   from scipy.linalg import solve, eigh
            A = np.random.random(9).reshape(3,3)
            A = A + np.transpose(A)      # form a symmetric matrix
            A
```

Out[120]:   array([[1.71865176, 0.78440106, 1.0647321],
                   [0.78440106, 0.70674838, 1.21531498],
                   [1.0647321 , 1.21531498, 1.15157901]])

We build the matrix $\mathbf{A}$ by generating 9 random numbers and reshape them into a 3×3 matrix. Then we add its transpose so the result is a symmetric matrix. The latter is necessary for real eigenvalues later on.

The column matrix $\mathbf{c}$ is prepared the same way,

```
In [121]:   c = np.random.random(3)
            c
```

Out[121]:   array([0.50965419, 0.95224918, 0.47953584])

The solution is simply obtained with,

```
In [122]:   x = solve(A, c)
            x
```

Out[122]:   array([-0.21045712, -0.65083933, 1.29786239])

To verify the solution, we substitute it back into the linear system,

```
In [123]:   np.dot(A, x)
```

Out[123]:   array([0.50965419, 0.95224918, 0.47953584])

As stated in Sect. 3.3.2, the dot operation works like matrix multiplication for 2D arrays. We see the resultant is equal to $\mathbf{c}$.

The eigenvalue and eigenvectors (see Sect. 6.2.3) may be obtained with,

```
In [124]:   vals, vecs = eigh(A)
            vals, vecs
```

---

[11] This is needed to find the quantum bound state energies of a particle in a spherical well in 3D.

```
Out[124]:  (array([-0.31107143,   0.60686089,   3.28118968]),
             array([[-0.05962799,   0.76124704,  -0.64571468],
                    [-0.7460073 ,  -0.46379005,  -0.47788271],
                    [ 0.66326285,  -0.45321268,  -0.59555072]]))
```

We have three eigenvalues, and three eigenvectors. The latter are stored in columns, for example,

```
In  [125]:  vecs[:,0]
```

```
Out[125]:  array([-0.05962799, -0.7460073 ,  0.66326285])
```

This is the first eigenvector. It is normalized,

```
In  [126]:  np.sum(vecs[:,0]**2)
```

```
Out[126]:  0.9999999999999997
```

It is also orthogonal to other eigenvectors,

```
In  [127]:  np.dot(vecs[:,0], vecs[:,1])
```

```
Out[127]:  -2.220446049250313e-16
```

It is practically zero (though not mathematically), but it is the limit of double precision seen earlier.

3.6   Speed Gain with Numba

It happens to every programmer at some point that we choose a robust algorithm for a problem and develop a code that works well in every aspect but it is just too slow. What to do? Code optimization, of course. Python has this undeserved perception that it is slow. It can be, but it does not have to be. For instance, explicit, nested looping will be slow. So we should avoid that as much as possible, by replacing explicit looping with Numpy's implicit looping (Sect. 3.3.2), for example. To the extent that that alone does not solve the speed problem, we can optimize in other ways.

That is where Numba comes in. Numba[12] can optimize a function with just-in-time compilation to speed up the execution. Before using Numba, we need to know where the speed bump is. Test the part of the code that you think is causing the slowdown. For instance, nested loops are an obvious target but it could be something else. Isolate that part, and time its execution with the system timer (see Program A.1.8). Once we are sure, put that part in a function and use @jit command of Numba to compile that function on the fly.

Let us illustrate it with an example on wave propagation (Sect. 6.2). A simplified wave model can be written as (see Eq. (6.21)),

$$u_n^{t_2} = u_{n-1}^{t_1} + u_{n+1}^{t_1} - u_n^{t_0}, \tag{3.2}$$

---

[12] https://numba.pydata.org.

where u_n^t denotes the wave amplitude at grid n and time t. Equation (3.2) advances the amplitudes to a later time t_2 from the amplitudes at earlier times t_1 and t_0. It can be coded easily with explicit looping,

```
def calc(u0, u1):
    u2 = - u0
    for n in range(1, N-1):
        u2[n] += u1[n-1] + u1[n+1]        # left, right
    return u2
```

In a simple program like Program A.1.8, the function `calc` is the bottleneck. For many grid points or time steps, explicit looping can take a lot of time.

Now, we can just put `@jit` immediately above `calc` as

```
@jit
def calc(u0, u1):
    ... same ...
```

Technically, the function `calc` is said to be decorated and compiled at runtime, all done without our doing anything. With this simple, almost magic trick, we get a speedup factor of 12 to 15 on average. If you have a working but slow program, try Numba optimization first. It may be all you need.

3.7   Exercises

3.1. Interactive plotting of free fall motion. Consider free fall motion with the position and velocity given by $y = v_0 t - \frac{1}{2}gt^2$ and $v = v_0 - gt$ (see Sect. 2.5.2).

   (a) Modify Program A.1.1 to interactively plot the position y or velocity v as a function of time t using IPywidgets and `interactive`. Now the slider k should be renamed v0 that returns the initial velocity v_0 in the range $[-5, 5]$ m/s, and the selection menu `func` should select either the position or the velocity. The time range should be $[0, 1]$ s.

  ***(b)** Do the same but with Python widgets as in Program A.1.2.

3.2. VPython animation. Program A.1.4 shows a sphere oscillating about the origin with the position $x = 2\sin(2t)$. Let us add an arrow to indicate its velocity as well.

  (a) Create an arrow representing the velocity vector after the box in Program A.1.4 with the following,

```
velvec = vp.arrow(axis=vp.vector(2,0,0))   # axis = arrow direction
```

In the loop, update the position of the arrow to that of the ball,

```
velvec.pos = ball.pos
```

You should now see the velocity vector moving with the ball.

(b) As the ball accelerates or changes direction, you will notice that the velocity vector stays the same. The velocity is given by $v = dx/dt = 4\cos(2t)$. To make the arrow proportional to the velocity, you will need to update its axis accordingly,

```
velvec.axis = vp.vector(2*vp.cos(2*t), 0, 0)
```

We use an amplitude of 2 so the arrow is not too big. Describe what you observed.

*(c) Further improvement of the visualization is possible. You may notice that the arrow becomes invisible, hidden inside the sphere when it slows down. To remedy the situation, you can either make the minimum length of the arrow to be the radius of the sphere, or put it in front of the sphere. Implement these strategies and comment on the results.

3.3. Manual improvement of numerical calculations. There are times when even the most sophisticated numerical computations need a little human help. A case in point is the integral $I = \int_0^\infty dx \, \sin^2(x)/x^2 = \pi/2$ encountered earlier (Sect. 3.5). The integral converges sufficiently slow that straight numerical integration yields low accuracy. The problem here is due to the combination of a slowly decaying integrand and the upper limit of the integral being infinity, and it can be hard to deal with numerical infinity. We will address this with manual intervention.

(a) Investigate the convergence first. Break up the interval of integration into smaller pieces of size a each, and add up the contributions. Let $f(x) = \sin^2(x)/x^2$ and $I_n = \int_{na}^{na+a} f(x)\,dx$, with $n = 0$ to m, so $I \simeq \sum I_n$. Calculate I_n with quad (see Sect. 3.5, Program A.1.7), choose $a = 100$ and $m = 4$. Compare the numerical and exact results. How many digits of accuracy is obtained? Increase m to 10, 20, and describe how fast, or slow, the result I converges.

(b) To improve convergence, first split the integral into two parts, $I = \int_0^a f(x)\,dx + \int_a^\infty f(x)\,dx = I_0 + I_r$. Show that the remainder can be written as $I_r = 1/2a - \frac{1}{2}\int_a^\infty dx \, \cos(2x)/x^2$. Break up the latter integral into subintervals of a as in part 3.3a above, and sum up the the contributions for $m = 4$. Add I_r and I_0 to obtain the full integral I. Describe the convergence compares with part 3.3a. Also try a few different m values, e.g., 10, 20 for comparison. Does it converge faster with the same upper limit?

Explain or speculate why.

3.4. Wien's displacement law. We solved for Wien's displacement law in the frequency domain in Sect. 3.3 analytically and again in Sect. 3.5 numerically. Let us express it in the wavelength domain, where the radiation spectrum is $f(x) = x^{-5}/(\exp(x) - 1)$, $x = hc/\lambda kT$.

(a) Solve for the maximum x_m with Sympy in terms of the Lambert W function. Express the result in the form $\lambda T = C$, and show that $C = 2.9 \times 10^{-3}$ m·K.

(b) Do the same in Scipy with `fsolve`. You can use the defining equation obtained with Sympy above.

*(c) Optionally solve the defining equation with the bisection method `bisect` in Scipy. Describe how you chose the root bracket.

3.5. Calculation of expansion coefficients. Let the wave function in the infinite potential well be $\psi(x) = Ax^2(a - x)$, $0 \leq x \leq a$. Carry out the following calculations with Sympy.

(a) Find the normalization constant A such that $\int_0^a |\psi(x)|^2 \, dx = 1$.

(b) Make a prediction whether you expect the expansion coefficient c_n to be zero for even n as was the case discussed in Sect. 3.3 (see Eq. (3.1) and also Sect. 7.3.1). Write down your prediction.

(c) Calculate c_n and the sum $\sum_n c_n^2$ (see Program A.1.5). Explain the results and compare them with your prediction.

*(d) Repeat the above with Scipy (see Program A.1.7) for $1 \leq n \leq 10$. What is the dominant term, why? Also use Numpy functions for sine and square roots first, then replace them with Sympy ones (import the latter, of course). Optionally calculate $\sum_n c_n^2$ for the first 10 terms. Discuss your observations.

3.6. Fourier transform.

(a) Given $\psi(x) = A \exp(-a|x|)$, $a > 0$, compute its Fourier transform defined as $\phi(k) = (2\pi)^{-1/2} \int_{-\infty}^{\infty} \psi(x) \exp(-ikx) \, dx$. Help Sympy with specific assumptions such as a is positive and k real. The result $\phi(k)$ is known as a Lorentzian. Graph $\phi(k)$.

(b) If $\psi(x)$ represents a wave function in position space, then $\phi(k)$ is the wave function in momentum space, and can also be regarded as an expansion coefficient like c_n in Exercise 3.5 but in the continuous variable k. Show that if $\psi(x)$ is normalized, so is $\phi(k)$, namely $\int_{-\infty}^{\infty} \phi^2(k) \, dk = 1$.

3.7. SHO in the forbidden region. Calculate the probability of the simple harmonic oscillator in the first excited state $n = 1$ in the classically forbidden region. Is the probability larger or smaller than that of the ground state (Sect. 3.3.2)? Explain.

References

M. Abramowitz and I. A. Stegun. *Handbook of mathematical functions: with formulas, graphs, and mathematical tables*. (Dover, New York), 1970.

M. L. Boas. *Mathematical methods in the physical sciences*. (Wiley, New York), 3rd edition, 2006.

I. S. Gradshteyn and I. M. Ryzhik. *Table of integrals, series, and products*. (Academic Press, New York), 1980. Translated and edited by A. Jeffrey.

Basic Algorithms

4

Symbolic computation discussed in the last chapter is an important component in scientific computing where numerical methods are also essential. We discuss select, elementary algorithms that are basic building blocks commonly used in scientific computing. We briefly introduce each algorithm mainly for quick reference. Applications and more detailed discussions where appropriate can be found in the next chapters focusing on specific problems.

4.1 Finite Difference

The basic idea of finite difference is to replace infinitesimals in calculus with chunks of finite size in numerical computation. Let us suppose we have a function $f(x_n)$ known at discrete points x_n, $n = 0, 1, 2, \ldots$ is an integer. Unless noted otherwise, the points x_n are assumed to be equally spaced, but this does not have to be the case in general. The interval is usually denoted by Δx, then $x_{n+1} = x_n + \Delta x$, and often it is useful to denote $f_n \equiv f(x_n)$.

Numerical Differentiation

First Derivatives

We associate the first derivative with the slope of a function at a single point. However, because a slope indicates a trend, we cannot obtain the first derivative if the function is known only at a *single point*. In the absence of a function in terms of a continuous variable, numerical first derivatives require at least *two points*.

Given $f(x_n)$ and $f(x_n + \Delta x)$, the slope is approximated by the straight line between the two points, and the first derivative is

© The Author(s), under exclusive license to Springer Nature Switzerland AG 2023
J. Wang and A. Wang, *Introduction to Computation in Physical Sciences*, Synthesis Lectures on Computation and Analytics,
https://doi.org/10.1007/978-3-031-17646-3_4

$$f' = \frac{f(x_n + \Delta x) - f(x_n)}{\Delta x} = \frac{f_{n+1} - f_n}{\Delta x}. \tag{4.1}$$

This is a first-order result. For example, if $x = x(t)$ is the position of a particle as a function of t (the independent variable), the velocity is given by the first derivative, $v = dx/dt \simeq \Delta x/\Delta t$ with $\Delta x = x_{n+1} - x_n$ (see Eq. (5.1)).

We usually think of f' as the first derivative at x_n, but it could equally be that at x_{n+1}. This ambiguity cannot be resolved with only two point, and in fact it can be used to our advantage to obtain a better approximation next.

Improved First and Second Derivatives

To obtain improved first derivatives as well as second or higher derivatives with an estimate of local error, let us begin with the Taylor series around x to third order,

$$f(x \pm \Delta x) = f(x) \pm f'\Delta x + \frac{1}{2}f''(\Delta x)^2 \pm \frac{1}{3!}f'''(\Delta x)^3 + O((\Delta x)^4). \tag{4.2}$$

Now we can find f' with a cancellation trick by subtracting both sides of Eq. (4.2) simultaneously,

$$f(x + \Delta x) - f(x - \Delta x) = 2f'\Delta x + O((\Delta x)^3),$$

so even terms cancel out, and the local error is of the order $O((\Delta x)^3)$. Dividing out $2\Delta x$, we have

$$f'(x) = \frac{f(x + \Delta x) - f(x - \Delta x)}{2\Delta x} + O((\Delta x)^2), \quad \text{or} \quad f_n' = \frac{f_{n+1} - f_{n-1}}{2\Delta x}. \tag{4.3}$$

Relative to Eq. (4.1), this result is second-order accurate, and has no ambiguity as to what point it refers. This three-point formula is just the average of $f'(x_{n+1})$ and $f'(x_{n-1})$ from Eq. (4.1), and so appropriately named the midpoint approximation.

For the second derivative, we add both sides of Eq. (4.2) to cancel out odd terms to obtain f'',

$$f''(x) = \frac{f(x - \Delta x) - 2f(x) + f(x + \Delta x)}{(\Delta x)^2} + O((\Delta x)^2), \quad \text{or} \quad f_n'' = \frac{f_{n-1} - 2f_n + f_{n+1}}{(\Delta x)^2}. \tag{4.4}$$

The above results is also accurate to second order.

Again, given the position as a function of time $x(t)$, its second derivative is the acceleration, $a = d^2x/dt^2 \simeq (x_{n-1} - 2x_n + x_{n+1})/(\Delta t)^2$. Actually, it is precisely this formula that is used in the wave equations, Eqs. (6.16), (6.19a) and (6.19b). One could continue the process to obtain higher derivatives, see Abramowitz and Stegun (1970).

4.2   Numerical Integration

Numerical integration, the inverse operation of numerical differentiation, is often necessary in scientific computing. Given a function $f(x)$, the goal of numerical integration is to find the integral,

$$I = \int_a^b f(x)\, dx. \tag{4.5}$$

Of course, we are assuming the integral is usually not known analytically, except for other purposes such as for testing. It is a good idea to check whether analytic solutions exist via tables of integrals (see Gradshteyn and Ryzhik 1980) or algebraic packages such as Sympy (Sect. 3.3).

The integrand $f(x)$ may be continuous, but can also exist as a set of data points. Either way, we wish to find the approximate area under the curve efficiently and accurately. The trapezoid method is a simple yet illustrative process.

Suppose the interval $[a, b]$ is divided into N, equidistant subintervals, so $\Delta x = (b - a)/N$. If we connect the adjacent points f_n by a straight line, each subinterval is a trapezoid, with an area given by $(f_n + f_{n+1}) \times \Delta x/2$. If we add up up all N subintervals, we have the total integral

$$I = \left(\frac{f_0}{2} + f_1 + f_2 + \cdots + f_{N-1} + \frac{f_N}{2} \right) \times \Delta x. \quad \text{(Trapezoid rule)} \tag{4.6}$$

Because the interior points are counted twice in the summation, the coefficients (weights) in front of f_1 to f_{N-1} are 1.

An easy improvement can be made to the trapezoid method. Instead of straight lines between two adjacent points, a parabola is chosen to fit the three points over a pair of neighboring subintervals. The area under the parabola between x_{n-1} and x_{n+1} is $(f_{n-1} + 4f_n + f_{n+1}) \times \Delta x/3$. This is the Simpson method. Now adding up all pairs, and counting the edges of each pair twice, the weights would alternate between 4/3 and 2/3 for the interior points, yielding a total integral as

$$I = (f_0 + 4f_1 + 2f_2 + \cdots + 2f_{N-2} + 4f_{N-1} + f_N) \times \frac{\Delta x}{3}. \quad \text{(Simpson rule)} \tag{4.7}$$

For the Simpson method, the number of subintervals N should be even because it works in pairs.

In practice, the Simpson method is efficient and accurate for smooth integrands, provided N is sufficiently large. It is especially useful for functions that are only piece-wise continuous, or in the case of discrete data points such the datasets from simulations or measurements.

One can adjust N to control accuracy of the Simpson method. For oscillatory functions and better accuracy control, more efficient integration schemes would be preferred. A general-purpose integration routine is offered via Scipy (Sect. 3.5), called quad (quadrature, a math jargon for numerical integration). It also has a standard `simps` routine, of course. To use `quad`, supply a user-defined function and the integration limits.

```
In [1]:  import numpy as np
         from scipy.integrate import quad, simps

         def psi(x):
             return x*(a-x)

         a = 1.0
         quad(psi, 0, a)[0]    # [0] contains integral
```

Out[1]: 0.16666666666666669

Because `quad` returns two values in a list, the integral and an error estimate, we select the first element only, the integral value.

The Simpson's method is very handy for a variety of tasks on the fly. To use `simpson`, simply provide the x and f data. For example, in quantum mechanics we often need to calculate normalization and expectation values given a wave function $\psi(x)$, such as $\langle x \rangle = \int |\psi|^2 x \, dx$, $\langle x^2 \rangle = \int |\psi|^2 x^2 \, dx$ and the like. We can use Simpson's method as follows.

```
In [2]:  x = np.linspace(0, a, 21)
         wf = psi(x)
         norm = simps(wf**2, x)
         simps(wf**2*x, x)/norm, simps((wf*x)**2, x)/norm
```

Out[2]: (0.5000000000000001, 0.28573191920201996)

The wave function $\psi(x) = x(a - x)$ is assumed to be within $[0, a]$ (infinite potential well, see Sect. 7.3.1). We used 20 subintervals with a grid size 0.05, which appears adequate in accuracy judging from the closeness of the results to the exact values, $\langle x \rangle = a/2$ and $\langle x^2 \rangle = 2a^2/7$ (see Exercise 4.6). Because it is not normalized, we need to divide the averages by the normalization `norm`. Note $\langle x^2 \rangle > \langle x \rangle^2$, so the uncertainty defined as $\Delta x = \sqrt{\langle x^2 \rangle - \langle x \rangle^2}$ is nonzero. The process requires us to supply only the wave function. Subsequent operations manipulate the data points only, and no new function needs to be defined.

4.3   ODE Solvers

Ordinary differential equations (ODEs) are common in science and engineering problems. For instance, to describe the motion of a particle (Chap. 5), we need a pair of ODEs like Eq. (5.7) for the position $x(t)$ and velocity $v(t)$,

$$\frac{dx}{dt} = v, \quad \frac{dv}{dt} = a. \tag{4.8}$$

Equation (4.8) is a typical system of coupled ODEs because the changes in position and velocity are interdependent. If we know the net force acting on a particle, we will know the acceleration a. The trajectory (solution) can be determined from Eq. (4.8) once the initial position and velocity are given. Numerical methods for solving ODEs are discussed in context to specific problems in Chap. 5. Here, we only outline the general idea for quick reference.

4.3.1 Euler's Method

The solutions to a set of coupled ODEs is basically an initial-value problem. It is intuitive to understand that a direct method would be to march forward one step at a time, starting from the initial conditions. Equation (4.1) hints at a straightforward time-stepping scheme. For example, applying Eq. (4.1) to $dx/dt = v$ of Eq. (4.8), and making appropriate substitutions $(f \to x, \; x \to t, \; f' \to v)$, we have for position stepping, $v = (x_{n+1} - x_n)/\Delta t$. We can similarly find velocity stepping to be $a = (v_{n+1} - v_n)/\Delta t$. Expressing the end-of-step values in terms of the start-of-step values gives

$$x_{n+1} = x_n + v_n \Delta t, \quad v_{n+1} = v_n + a_n \Delta t. \quad \text{(Euler's method)} \tag{4.9}$$

Equation (4.9) is the Euler's method, the simplest way to move forward one step. If a is nonzero, velocity will change between steps. That is why we have used v_n instead of just v in the position stepping in Eq. (4.9), in keeping with forward time-stepping. Similarly, if acceleration changes with time, we need to use $a_n = a(t_n)$. By combining both the position and velocity stepping methods, we can solve any motion with changing velocity due to arbitrary acceleration (see Eq. (5.6), Sect. 5.2).

4.3.2 Euler-Cromer Method

Sometimes the acceleration may not depend on time explicitly but implicitly via position. This is the case for the simple harmonic oscillator (Sect. 5.5.1) where the acceleration is a function of position $a(x)$ because the spring force is position dependent. Then, consistency would seem to suggest that we use $a(x_n)$ in Eq. (4.9). It turns out that would be wrong. Actual results show the correct way is to use $a(x_{n+1})$ instead,

$$x_{n+1} = x_n + v_n \Delta t, \quad v_{n+1} = v_n + a(x_{n+1}) \Delta t. \quad \text{(Euler-Cromer method)} \tag{4.10}$$

This modification, known as the Euler-Cromer method, effectively uses the latest available position in the calculation of acceleration. As later discussion in Sect. 5.5.1 shows, this method preserves important physical properties, including energy conservation. The Euler-Cromer method is as easy to use as standard Euler's method, and should be used in numerically solving conservative systems.

4.3.3 Leapfrog Method

The Euler-Cromer method is easy to use and works well enough for the most part in problems to be discussed later (Chap. 5). However, it is only accurate to first order. If we want to increase accuracy, the step size Δt will need to decrease, and the method becomes less efficient.

Luckily, there exists an easy way to make it more accurate and nearly as efficient, with only a slight tweak. The result is the leapfrog method (see Sect. 5.5.3). As the name suggests, the position and velocity updates are done at different times, leapfrogging each other.

Starting with Eq. (4.10), we first update the position for one half step to the midpoint x_m, then use it to advance the velocity for a full step, also using x_m to evaluate the acceleration at the midpoint. Finally another half step is taken to obtain the position. One full cycle is given as

$$x_m = x_0 + v_0 \times \frac{\Delta t}{2},$$
$$v_1 = v_0 + a(x_m) \times \Delta t,$$
$$x_1 = x_m + v_1 \times \frac{\Delta t}{2}. \quad \text{(Leapfrog method)} \tag{4.11}$$

At each stage, the method uses the latest position and velocity available. It is accurate to second order, significantly better than the Euler-Cromer method. This increased accuracy is achieved due to the calculation of acceleration at x_m, effectively the average of the initial and final positions, exactly the same reason as improved derivative Eq. (4.3) being more accurate.

The leapfrog method is almost as efficient because, even though Eq. (4.11) consists of three steps versus two for Eq. (4.10), usually the evaluation of acceleration is the most costly step, which is the same in both methods. For results to be comparable in accuracy, the leapfrog method is much faster, see Table 5.1 for performance comparison.

The idea of splitting the calculation of position in the leapfrog method may be used to our advantage in other situations. One example has to do with motion in cross electromagnetic fields where both the electric and the Lorentz forces act on an object simultaneously (Sect. 5.5.3). To simulate the motion correctly, it is necessary to calculate the effect of electric force in two half-steps like the position (see Eq. (5.24)).

Higher-Order Methods

There exist other higher-order ODE solvers. Among them, the most well-known are the Runge-Kutta family of methods. In fact, the Euler's method Eq. (4.9) is the first member of the Runge-Kutta family. Other members of the Runge-Kutta family are of second-, fourth-, and higher-order accuracy. One can access those higher-order methods with the Scipy function `odeint`, a general purpose ODE solver with error control.

4.4 Root Solvers

Given a function $f(x)$, there is a need sometimes to find values of x, where the function is zero $f(x) = 0$. Such values are called roots. Root solvers generally fall into two categories: methods that require a bracket and method that require only a seed.

The bisection method requires a bracket $[a, b]$ where a root is known to exist. A root is bracketed if the condition $f(a) \times f(b) < 0$ holds. The basic idea is: cut the bracket in half, find the half where the root lies, and repeat until the bracket is sufficiently small to precisely locate the root. The advantage of the bisection method is that it never fails to find a root if the initial bracket is valid. The disadvantage is that it is usually not the fastest root solver.

On the other hand, Newton's method is usually one of the fastest root solvers. It needs a seed value to get started. It then tries to project an approximate root at the intersection of the tangent line and the x axis. The process is repeated until it converges to a root. The success of the Newton's method depends on the behavior of function and the initial seed. When it works, it can be very fast. But unlike the bisection method, Newton's method can fail to find a root. So, unless speed is very important, one should almost always use the bisection method.

Scipy has several root solvers that we can use, `bisect` and `fsolve`. The former is the bisection method just discussed. The latter is a Newton-like method, and requires an initial seed to get started. The seed is a guess and does not have to be close to the actual root, though the closer the better. Here we use these two root solvers to find the allowed energies in a square well potential in quantum mechanics.

```
In [3]:  import numpy as np
         from scipy.optimize import bisect, fsolve

         def f(E):                      # allowed energy in square well
             s = np.sqrt(E + V0)
             return s*np.tan(np.sqrt(2)*s*a) - np.sqrt(-E)

         a = 2.0          # width
         V0 = 1.0         # depth

         bisect(f, -.99*V0, -.7*V0), fsolve(f, -.9*V0)
```

Out[3]: (-0.8342557172898887, [-0.83425572])

The potential generally supports both even and odd states, respectively referring to its symmetric or antisymmetric wave functions about the center. The function `f` defines the even state energy E satisfying the characteristic equation, $\sqrt{E + V_0} \tan(\sqrt{2(E + V_0)}a) - \sqrt{-E} = 0$, with V_0, a being the depth and width of the square well, respectively. It is written in atomic units where the energy unit is 27.2 eV and length unit 0.529 Å (see Sect. 7.3.3). The bracket is set to $[-0.99V_0, -0.7V_0]$ for the bisection method, because the root must be between the bottom and the top of the well. The initial guess is $-0.9V_0$ for the general root solver `fsolve`. Both results agree with each other. For this set of parameters, there is only one even state (the ground state). There is an odd state, however (see Exercise 4.6). Note that `fsolve` returns the roots in an array. It also allows for multiple seeds contained in an input array, and the output array will be equal in length. The roots, however, may not be unique because different seeds can converge to the same root.

4.5   Curve Fitting

In analyzing datasets from either measurements or simulations, sometimes we wish to fit the data points to a known model to extract the parameters, or to determine their trends and identify possible scaling behavior. Curve fitting is useful in this aspect because it can compute the best-fit parameters and could also help quantitatively distinguish one model from another.

To fit to a given model $f(x, a, b, \cdots)$, we compare how close the actual data points d_n are to the predictions $f_n = f(x_n, a, b, \cdots)$ of the model. The closeness is usually measured by the least-square error, defined to be $\Sigma = \sum(d_n - f_n)^2$, where for simplicity all data points are weighted equally. Now we minimize Σ to obtain the best-fit parameters.

Scipy has a `curve_fit` function that does all the leg work for us. An example using it is shown in Fig. 4.1. The data points show the position as a function of time in free fall generated with Program A.2.1. To "confuse" the fitting routine, random noise is added to the data.

Two models are used to test fit, one linear and the other quadratic. The quadratic model fits the data better than the linear model as expected. We can also obtain the relevant parameters such as the initial position, velocity, and gravitational acceleration g from the quadratic fit. With the level of noise, the best-fit gives $g = 9.3$ m/s^2.

Fig. 4.1 The position-time data in simulated free-fall. Random noise is added to the data points. A linear (dashed line) and a quadratic (solid line) fits are also shown

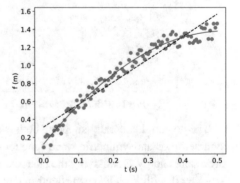

Locating Minima and Maxima

In data analysis, we often need to find the local minimum or maximum values, or extrema in general, in the data. Let y be a series of data points. The point y_i is a local extremum if it is either smaller or greater than its nearest (left and right) neighbors. We can write a function to test the conditions as follows (taken from Program A.2.2).

```
In [4]:  def minmax(y):       # return indices of minima and maxima
             mini = (y[1:-1]<=y[:-2])*(y[1:-1]<=y[2:])    # minima array
             maxi = (y[1:-1]>=y[:-2])*(y[1:-1]>=y[2:])    # maxima array
             mini = 1 + np.where(mini)[0]                 # indices of minima
             maxi = 1 + np.where(maxi)[0]                 # indices of maxima
             return mini, maxi
```

The function `minmax` takes a 1D input array and returns indices of minima and maxima as arrays in `mini` and `maxi`, respectively (see Program A.2.2 for an example using it). It checks only the interior points because end points have just one neighbor to one side so they cannot be extrema by definition.

Index slicing is ideally suited for this problem (see Sects. 2.4.2 and 3.3.2). The conditions $y_i \leq y_{i-1}$ and $y_i \leq y_{i+1}$ for local minimum values are checked with the product of `y[1:-1]<=y[:-2]` and `y[1:-1]<=y[2:]`, where `y[1:-1]` holds entries from the second to the second last elements, `y[:-2]` from the first to the third last, and `y[2:]` from the third to the last elements. All are truth arrays, and have the same length that is two elements fewer than the original array. Multiplication between two truth arrays is allowed and is equivalent to logical `and`. Addition of two truth arrays is also allowed, and is equivalent to logical `or` (Subtraction and division are also defined but not meaningful in the current context).[1] The conditions $y_i \geq y_{i-1}$ and $y_i \geq y_{i+1}$ for local maximum values can be checked similarly.

The function `np.where(mini)` returns the indices where the truth array is `True`. (`np.nonzero` would also work as `True` and `False` are internally represented by 1 and 0, respectively.) Because array `mini` begins at the second element relative to the data array `y`, we shift it up by 1 to obtain correct indices of minima. The process is analogous for maxima.

Program A.2.2 illustrates the process of automatically locating minima and maxima. Test data is generated from a decaying sine function $y = \exp(-x/2)\sin(2\pi x)$. The result is shown in Fig. 4.2. The method is robust and efficient. More applications can be found later (see Exercise 7.5).

---

[1] Conditional truth array indexing is known as advanced indexing, and is a powerful feature of Numpy. Basically, a conditional like `a < 0` returns a truth array where elements satisfying the conditional are set to `True`, and to `False` otherwise. The effect is that `b[a < 0]` selects only those elements of `b` at positions where the truth array is `True`.

Fig. 4.2 Automatic searching of local minima and maxima (Program A.2.2)

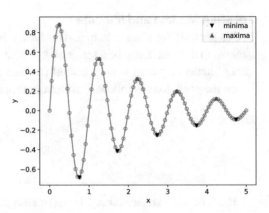

4.6   Exercises

4.1. Calculation of expectation values.

(a) Verify the exact values quoted in Sect. 4.2 for $\langle x \rangle = a/2$ and $\langle x^2 \rangle = 2a^2/7$ if the wave function is $\psi = x(a - x)$.

(b) Using the same wave function as in Exercise 3.5, calculate the expectation values $\langle x \rangle$ and $\langle x^2 \rangle$ with Simpson's method. Find the uncertainty Δx. Reuse the normalization constant A if you had already done Exercise 3.5. Otherwise, make sure to account for the normalization.

(c) Repeat the above calculation with the quad function. How do the results compare with those of Simpson's method? Was the grid size adequate?

(d) Calculate the expectation values of inverse powers of x, $\langle 1/x \rangle$ and $\langle 1/x^2 \rangle$. Compare with part 4.1b, and briefly comment on why $\langle 1/x^n \rangle \neq 1/\langle x^n \rangle$.

4.2. Discrete states in the square well potential.

(a) In Sect. 4.4, an example was given for finding the even state energy in a square well potential. There is another set of odd states with energy E satisfying the defining equation (in atomic units, used in this exercise), $g(E) = \sqrt{E + V_0} \cot \left[\sqrt{2(E + V_0)}\, a \right] + \sqrt{-E} = 0$. For the same parameters $a = 2$ and $V_0 = 1$, find the energy of an odd state. This should be the only odd state.

(b) Keeping $V_0 = 1$ fixed but increasing a, devise a method to find the smallest a such that a second odd state starts to appear.

*(c) Now, keeping $a = 2$ fixed but increasing V_0, find the smallest V_0 such that a second odd state barely exists. Proportionally, is it more effective to increase a or V_0 for the second odd state to appear?

4.3. Energy conserving integrator. Both the Euler and Euler-Cromer methods are first-order differential equation integrators (Sect. 4.3). But there is a big difference between them, namely: if the energy of a system is conserved, the Euler-Cromer method keeps the energy constant (within its first-order accuracy, of course), and the Euler method does not. We will investigate this next. The classical simple harmonic oscillator (SHO) in 1D can be described by the differential equations (similar to 3D, Eq. (5.14)),

$$\frac{dx}{dt} = v, \quad \frac{dv}{dt} = -\omega^2 x. \tag{4.12}$$

(a) Solve Eq. (4.12) with the Euler method Eq. (4.9). Initialize Δt, ω, x_0 and v_0 (e.g., 0.1, 1, 1, 0, respectively), then enter the main loop to step forward in time,

```
In [7]:  while t<20:
             x = x0 + v0 * dt
             v = v0 - x0 * dt
             ......
             x0, v0 = x, v
             t = t + dt
```

Record the position and velocity inside the loop. Plot the $x - t$ and $v - t$ curves. Describe the change of the amplitudes of x and v with time.

(b) Add a statement in the loop to record the energy as $E = \frac{1}{2}m\omega^2 x^2 + \frac{1}{2}mv^2$ (assuming $m = 1$). Plot E as a function of time. Describe the change.

(c) Next use the Euler-Cromer method Eq. (4.10). The only change would be in updating the velocity,

```
In [8]:  while t<20:
             x = x0 + v0 * dt
             v = v0 - x  * dt
             ......
```

Again graph the $x - t$ and $v - t$ curves. How do the amplitudes of x and v change with time? Compare with the Euler method.

(d) Finally, plot the energy as a function of time. Does the energy stay bounded? You may notice that there are small oscillations around the initial energy. How can we reduce them?

References

M. Abramowitz and I. A. Stegun. *Handbook of mathematical functions: with formulas, graphs, and mathematical tables*. (Dover, New York), 1970.

I. S. Gradshteyn and I. M. Ryzhik. *Table of integrals, series, and products*. (Academic Press, New York), 1980. Translated and edited by A. Jeffrey.

Force and Motion

5

In the first part of discussions up to now, we have introduced basic elements of programming and toolboxes for scientific computing. Our goal is not computation in and of itself, although it is an important and necessary component. Rather, we aim to solve physical and engineering problems with computational methods. In the second part starting from this chapter, we consider applications of computational techniques in the context of specific problems ranging from classical to quantum mechanical, and from simple to complex systems.

5.1 Motion with Constant Velocity

Our physical world is constantly in motion, and the study of motion is called kinematics. We begin with the simplest motion described by Newton's first law: in the absence of a net external force, an object stays in motion with constant speed and direction.

5.1.1 Kinematic Concepts

We can think of the object as a point particle, having a well-defined position x at time t (the assumption of a well-defined position is challenged in quantum mechanics, see Chap. 7). Under Newton's first law, it necessarily follows that the particle will move along a straight line, which we regard as the x-axis. To follow the motion, we can mark the positions of the particles on the x-axis as position-time pairs, namely x_0 at t_0, x_1 at t_1, and so on, with the usual convention of the initial condition starting from x_0 and t_0.

© The Author(s), under exclusive license to Springer Nature Switzerland AG 2023
J. Wang and A. Wang, *Introduction to Computation in Physical Sciences*, Synthesis Lectures on Computation and Analytics.
https://doi.org/10.1007/978-3-031-17646-3_5

What questions could be asked and what concepts developed from those data points? The most direct question we might ask is this: how fast will the particle move from one position to another, say x_0 to x_1? The answer depends on two factors: how far apart the positions are, $\Delta x = x_1 - x_0$ (distance, or more precisely, displacement), and how long the time interval is, $\Delta t = t_1 - t_0$. Within a given time interval, the larger the displacement, the faster the motion; and within a given displacement, the smaller the time interval, the faster the motion. This motivates us to develop a "fastness" concept, v, to measure this as a ratio

$$v = \frac{\Delta x}{\Delta t}. \tag{5.1}$$

We recognize this "fastness" thus defined is just the concept of velocity we first encounter in kinematics. Either a larger Δx or a smaller Δt would yield a larger v, meaning faster motion as is commonly perceived. Note, however, velocity v can be positive or negative, carrying the same sign as Δx. If we strip the sign information, we obtain speed $|v|$. This careful albeit straightforward exercise shows that such concepts are constructs purposefully formulated to characterize motion, and not simply found in nature the way objects are. Other concepts we encounter going forward such as acceleration (Sect. 5.2) are constructed similarly.

5.1.2   Motion with Finite Time-Stepping

Let us return to motion with constant velocity described by Newton's first law. Given velocity v in Eq. (5.1), we can predict the change of position Δx after time interval Δt as

$$\Delta x = v \times \Delta t. \tag{5.2}$$

Though Eq. (5.2) is equivalent to Eq. (5.1), it implies an important concept of time stepping, namely, given the position x_0 at t_0, we can obtain the position x_1 at t_1 as $x_1 = x_0 + v \times \Delta t$. Subsequent positions x_n may be calculated by stepping forward from the previous position, one step at a time, assuming equal time interval:

$$x_1 = x_0 + v \times \Delta t,$$
$$x_2 = x_1 + v \times \Delta t,$$
$$\ldots \ldots$$
$$x_n = x_{n-1} + v \times \Delta t. \tag{5.3}$$

The forward-stepping method Eq. (5.3) is also called the Euler method in numerical computation, the simplest method for solving ordinary differential equations. The discussion here provides context to time stepping first described in Sect. 2.12.1.

How do we track the motion computationally? Admittedly it is vastly over-complicating matters by invoking computation for such a simple motion with constant velocity. However, this is useful precisely because the simplified problem allows us to understand the method (algorithm) clearly, test it against known results, and extend to nontrivial cases as we will see shortly. It is a proven successful practice to start with simple physical models and extend them to more complex situations.

Now is a good time for us to model the time-stepping process. First let us use the Jupyter notebook as a calculator, stepping forward once as follows.

```
In [1]: x0 = 0.0
        v = 2.0
        dt = 0.1
        x1 = x0 + v*dt
        print(x1)
```

The code introduces three variables v, x_0 and Δt and assigns arbitrary values (SI units are assumed unless stated otherwise). Then x_1 is calculated according to Eq. (5.3), and the result is printed. Make sure the result is as expected. Should an error occur, check the variable names and other issues like statement alignment. See Sect. 2.11 for finding and fixing errors.

Usually we are interested in a series of data points. Suppose we wish to move n steps, the code to accomplish that is logically via a loop.

```
In [2]: x0 = 0.0
        v = 2.0
        dt = 0.1
        for i in range(10):
            x1 = x0 + v*dt
            print(x1)
            x0 = x1
```

The initial assignment is the same as before. We have added a `for loop`, iterating $n = 10$ steps. In each step, we calculate the position the same way as in the single-step code, print the values, and most important, reseed the old value `x0` with the new `x1` in preparation for the next step. The loop block ends after the last statement within the same indentation.

Again, make sure to check the results are as expected. You will notice that scanning a lot of numbers is not the best way to spot trends or patterns. This is where graphics and visualization is important.

Let us now graph the results. To do so, we extend the code as follows.

Program 5.1 Motion with constant velocity.

```
 1  import matplotlib.pyplot as plt
    x0 = 0.0
 3  v = 2.0
    dt = 0.1
 5  t = 0.0
    tlist = []
 7  xlist = []
    for i in range(10):
 9      tlist.append(t)
        xlist.append(x0)
11      x0 = x0 + v*dt
        t = t + dt
13
    plt.plot(tlist, xlist)
15  plt.xlabel('time (s)')
    plt.ylabel('x (m)')
17  plt.show()
```

We first import the standard plotting library `matplotlib` and give it the common alias `plt` for easy reference later (see Sect. 2.10). We introduce a new variable t to keep track of time, and two lists, one for time and one for position, needed to store the results for plotting. In the loop, the current time and position are added to the respective lists using the `append` method built into Python. Next we update the position and time. Note that we get rid of x1 used earlier. We can do this because the assignment operator = in programming is not an equality statement in the mathematical sense. The statement takes the value of x0, add v*dt to it, and assigns the result back to x0, such that the variable x0 has changed its value after execution of the statement. This way we avoid having to reseed x0 like before. Of course we could have done the same thing before as well, but for clarity we chose to use both.

After the loop, we plot x verses t from stored values in the lists, add axis labels and units that proper figures should have. The last line `plt.show()` displays the composed figure in memory (see Sect. 2.10. It may not be needed depending on your environment, e.g., Jupyter notebooks do not need it).

Now we have built a fully functional program complete with visualization. We should see a straight line showing the linear increase of position with time. We will extend the program to motion with constant acceleration next. However, we can play what-if scenarios with different initial conditions. For instance, what effect would we expect if we change the initial position? The velocity? Make a prediction what would happen if the velocity is negative; sketch the predicted x vs. t curve; run the program with a negative velocity (say -2.0) and compare the results with your prediction (also see Exercise 5.1).

5.2   Modeling Motion with Changing Velocity

Motion with constant velocity is a special case of Newton's first law. Generally, we expect the particle to speed up or slow down with changing velocity. How do we characterize the change of velocity? Similar to change of position discussed earlier, we can record change of velocity by setting up an axis for velocity, marking its values v_0 at t_0, v_1 at t_1, and so on. Just as we asked how fast a particle goes from one point to another, we could ask how quickly velocity changes from one value to another on the *velocity* axis. Again, the same reasoning leads us to consider $\Delta v = v_1 - v_0$ and $\Delta t = t_1 - t_0$: the larger the Δv for a given Δt or the smaller the Δt for a given Δv, the quicker the change. Consequently, we develop a concept to measure this as a ratio a

$$a = \frac{\Delta v}{\Delta t}. \tag{5.4}$$

Equation (5.4) tells us the rate of change of velocity, namely, acceleration. Like velocity, acceleration also carries a sign, the same sign as Δv, and measures the change of velocity, not of speed. So it is possible for a particle to either speed up or slow down with a positive acceleration. For example, changing velocity from 0 to 10 m/s or from -10 m/s to 0 over some Δt leads to the same positive acceleration, but the motion is speeding up in one case and slowing down in the other.

Newton's second law governs directly how velocity changes with $a = F/m$, given the net force F acting on a particle of mass m. Knowing a, we are able to follow change of velocity in the same parallel way to change of position,

$$\Delta v = a \times \Delta t. \tag{5.5}$$

5.2.1   Dynamic Equations of Motion

Both Eqs. (5.2) and (5.5) stand on their own merit, but they present a very powerful combination that each alone would not achieve: the combination allows us to extend motion from constant to arbitrary motion with changing velocity. We already know how to update position given v, now we can also update velocity in the same way given a. This way both position and velocity can be updated simultaneously, and we can describe any dynamic motion when the two equations are coupled.

The recipe for the extension is similar to Eq. (5.3):

$$x_1 = x_0 + v_0 \times \Delta t, \quad v_1 = v_0 + a \times \Delta t,$$
$$x_2 = x_1 + v_1 \times \Delta t, \quad v_2 = v_1 + a \times \Delta t,$$
$$\cdots \cdots$$
$$x_n = x_{n-1} + v_{n-1} \times \Delta t, \quad v_n = v_{n-1} + a \times \Delta t. \tag{5.6}$$

Now we have a powerful method at our disposal with Eq. (5.6), and let us take it for a spin. Again, consider the simplest motion with changing velocity: motion with constant nonzero acceleration a, such as in free fall. Velocity will change at a constant rate. We aim to extend Program 5.1 in this case as follows.

Program 5.2 Motion with constant acceleration.

```
 1 import matplotlib.pyplot as plt
   a = 1.0
 3 v = 2.0
   x = 0.0
 5 dt = 0.1
   t = 0.0
 7 tlist = []
   xlist = []
 9 vlist = []
   for i in range(20):
11     tlist.append(t)
       xlist.append(x)
13     vlist.append(v)
       x = x + v*dt
15     v = v + a*dt
       t = t + dt
17
   plt.plot(tlist, xlist)
19 plt.xlabel('time (s)')
   plt.ylabel('x (m)')
21 plt.figure()
   plt.plot(tlist, vlist)
23 plt.xlabel('time (s)')
   plt.ylabel('v (m/s)')
25 plt.show()
```

This program is very similar to the previous one with constant velocity with several additions: constant acceleration a, a list to store velocity, velocity update inside the loop, and an additional figure to plot velocity-time curve. Note that now position update requires current value of velocity, and velocity changes with each iteration. Though we assume a constant acceleration here, in many situations the force depends on the position, so acceleration, and hence velocity update, requires current value of position. The two variables are co-dependent, or coupled.

We often describe problems in physics and engineering in terms of ordinary differential equations (ODEs). In the limit of infinitesimal change $\lim \Delta t \to dt$, we can rewrite Eqs. (5.1) and (5.4) as a set of ODEs,

$$\frac{dx}{dt} = v, \quad \frac{dv}{dt} = a. \tag{5.7}$$

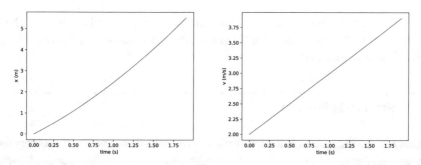

Fig. 5.1 Motion with constant acceleration. Position- and velocity-time curves, respectively

We speak of these two ODEs as coupled equations. The algorithm of Eq. (5.6) is the Euler's method for the ODEs, which is implemented in Program 5.2.

Figure 5.1 displays the results from Program 5.2. The position-time line is now curved rather than straight, a consequence of velocity changing with time. If we choose a time window (interval Δt) and slide it back and forth on the time axis, the change of position will be different with the same window, as expected. We also expect the amount of change (Δx) to vary depending on how curved the line is (i.e., the curvature). According to Eq. (5.1), velocity is the slope of the position-time curve (rise over run). The slope on the $x - t$ curve is becoming steeper in this case, so velocity must be increasing. Indeed, we see velocity increasing linearly with time, consistent with what we expect. The same reasoning from Eq. (5.4) leads us to the conclusion that acceleration is the slope of the velocity-time curve. Here the $v - t$ curve is a straight line, meaning a constant slope and constant acceleration, exactly what we assumed. To be precise, Eqs. (5.1) and (5.4) give average velocity and acceleration, and Eqs. (5.7) give instantaneous velocity and acceleration (see see Exercise 5.1).

Again one can run the program with different initial conditions to see the effects. For instance, what if the initial velocity is negative? what if acceleration is negative? In each case, make a prediction as to what the the position- and velocity-time curves look like, sketch them, then run the program to check your predictions (see Exercise 5.1).

From the discussion of the results, the program seems to be working as expected. In fact we should validate programs further against analytic solutions wherever possible. For motion with constant acceleration, they are available from introductory physics as

$$x = x_0 + v_0 t + \frac{1}{2}at^2, \quad v = v_0 + at. \tag{5.8}$$

The position x changes quadratically with time (a parabola) and velocity linearly with time. To confirm, we should plot the analytic results along with the numerical ones. We leave this task as an exercise (Exercise 5.1).

Now that step by step we have built and validated a program with Euler's method for 1D motion, can we extend it to 2D or 3D motion, as well as non-constant acceleration? We discuss these extensions next.

5.3   Projectile Motion

Arguably, projectile motion is the best well-known problem in physics. It is observable in daily life and one of the first real-world examples encountered in introductory physics. On the one hand, although projectile motion is multidimensional, it is readily solvable if air resistance is neglected. On the other hand, realistic projectile motion requires us to take into account air resistance, or drag. Because inclusion of drag forces renders the problem unsolvable analytically, we must rely on computation to solve it. Projectile motion is an ideal example to study computational techniques because it is relatively simple, intuitive, and its basic features are already familiar to us. Coupled with analytic solutions in 2D, we can readily test and validate those techniques in the limit of small air resistance where analytic solutions are a good approximation.

In ideal projectile motion, the particle moves under gravity directed vertically downward. Assuming x and y in the horizontal and vertical directions, respectively, the equations of motion Eq. (5.7) are

$$\frac{d\mathbf{r}}{dt} = \mathbf{v}, \quad \frac{d\mathbf{v}}{dt} = -g\hat{\jmath}. \tag{5.9}$$

These equations are straightforward generalizations from Eq. (5.7), with x, v, and a replaced by their respective vector equivalents as $\mathbf{r} = x\hat{\imath} + y\hat{\jmath}$, $\mathbf{v} = v_x\hat{\imath} + v_y\hat{\jmath}$, and $\mathbf{a} = -g\hat{\jmath}$, where $\hat{\imath}$ and $\hat{\jmath}$ are the usual unit vectors, and $g = 9.8$ m/s^2 the constant gravitational acceleration.

Correspondingly, the time-stepping method takes on the vector form $\Delta\mathbf{r} = \mathbf{v} \times \Delta t$ and $\Delta\mathbf{v} = \mathbf{a} \times \Delta t = -g \times \Delta t\hat{\jmath}$, and we can express the vector version of Eq. (5.6) as

$$\mathbf{r}_n = \mathbf{r}_{n-1} + \mathbf{v}_{n-1} \times \Delta t, \quad \mathbf{v}_n = \mathbf{v}_{n-1} - g \times \Delta t\hat{\jmath}. \tag{5.10}$$

Similarly, the previous program just needs to be extended via vector operations. Fortunately, this can be achieved simply with the powerful numpy support of vector array operations. The full program is given elsewhere (Program A.3.1). Structurally, it is materially the same as Program 5.2, but instead of scalars, we initialize acceleration, velocity and position as 2D numpy arrays:

```
In [3]: a = np.array([0.0, -g])
        r = np.array([0.0, 0.0])
        v = np.array([4.0, 9.0])
```

As discussed earlier (Chap. 2), unlike lists, these arrays support vector operations. Take this line for example,

```
In [4]: r = r + v*dt
```

Each element of velocity is multiplied by the time step and then added to position, again element-wise. It behaves effectively like real vector operations in $\mathbf{r}' = \mathbf{r} + \mathbf{v} \times \Delta t$ as follows:

$$\mathbf{r}' = (x + v_x \times \Delta t)\hat{\imath} + (y + v_y \times \Delta t)\hat{\jmath}. \tag{5.11}$$

After the calculation, we convert the lists to numpy arrays as

```
In [5]: rlist = np.array(rlist)
```

We do this so we can obtain the components via array slicing. The $x-$ and $y-$components are stored in the first and second columns, respectively, rlist[:,0] and rlist[:,1]. For example, the following graphs the $x-$ and $y-$components of position versus time,

```
In [6]: plt.plot(tlist, rlist[:,0], tlist, rlist[:,1])
```

Velocity can be graphed in the same way, and the full results (not shown here, see program output) should be very familiar to anyone who studied projectile motion from introductory physics. The horizontal position-time curve is a straight line because the horizontal velocity is constant due to the fact that gravitational acceleration is in the vertical direction.

As a result, the corresponding vertical position-time curve is a downward parabola, and the vertical velocity decreases linearly with its slope equal to $-g$. Note that at the top of the trajectory, namely, maximum $y-$position, v_y is zero but a_y is still $-g$, not zero as is sometimes misconstrued. This is because we cannot tell velocity looking at a single data point in position, and we need at least two points to determine the change in position and obtain velocity from Eq. (5.1). Likewise, we cannot determine acceleration with a single data point in velocity, and we must look at the trend of velocity between at least two points. In this case $v_y - t$ clearly shows a constant slope everywhere, hence constant acceleration everywhere, including at the top.

Though the program is working as expected, we still should compare the results with well-known analytic solutions. For example, we can test the accuracy of the trajectory and of energy conservation, and determine what factors affect the accuracy (see Exercise 5.2). We can also try virtual target practice (Exercise 5.3). Such validation is important because as we will discuss next realistic projectile motion involving drag has no known solutions.

5.4 Projectile Motion with Air Resistance

Earlier we discussed ideal projectile motion under gravity only. From everyday life, we know air resistance, or drag, affects motion in the air, as someone observing a feather flapping up and down can confirm. We can effectively model the force from air resistance empirically as a velocity dependent force (see Taylor (2005)),

$$\mathbf{F}_d = -b_1 \mathbf{v} - b_2 v \mathbf{v}. \tag{5.12}$$

In Eq. (5.12), $\mathbf{v}$ is the velocity of the projectile relative to air, $v = |\mathbf{v}| = \sqrt{v_x^2 + v_y^2}$ the speed, b_1 and b_2 are constants for the linear and quadratic drag forces. Note that the signs before the terms are negative indicating the drag force being opposite to the direction of velocity. Clearly there is no drag if $\mathbf{v} = 0$.

The linear term in Eq. (5.12) is a viscous force (sticky air molecules), and is important at low speeds when airflow is smooth (laminar flow). The origin of the quadratic force in Eq. (5.12) comes from a moving object displacing and accelerating air mass in front (turbulent flow). It is dominant practically at all but the lowest speeds. The coefficients b_1 and b_2 depend on the shape and surface properties of the object. Though both can be measured experimentally or modeled theoretically (Frohlich (1984)), we will assume nominal values for illustration purposes.

The acceleration due to the drag force Eq. (5.12) is $\mathbf{a}_d = \mathbf{F}_d/m = -c_1\mathbf{v} - c_2v\mathbf{v}$ with $c_{1,2} = b_{1,2}/m$, assuming mass m. The equations of motion can be modified from Eq. (5.9) to include this acceleration as

$$\frac{d\mathbf{r}}{dt} = \mathbf{v}, \quad \frac{d\mathbf{v}}{dt} = -g\hat{j} - c_1\mathbf{v} - c_2v\mathbf{v}. \tag{5.13}$$

Because of the quadratic drag force, these equations no longer admit closed-form analytic solutions, so we have to solve them numerically. But this presents no particular hurdle based on Program A.3.1. We only need to include $\mathbf{a}_d$ according to Eq. (5.13). The modified Program A.3.2 is structurally similar to Program A.3.1. The main difference is this:

```
In [7]:  speed = np.sqrt(v[0]**2 + v[1]**2)
         a = ag - c1*v - c2*speed*v
```

The acceleration now includes both linear and quadratic drag forces. The vector nature of the velocity array makes the calculation of acceleration straightforward. Note the main loop is also changed to a while-loop terminating once the y-position becomes negative instead of a fixed number of steps. We can regard the x-position at the end as the range.

The results from Program A.3.2 are shown in Fig. 5.2. We see immediately key differences from ideal projectile motion. The x position is curved instead of being straight. It bends downward, reflecting the fact the v_x is decreasing because the drag force induces a negative horizontal acceleration, as well as a vertical one, causing v_x to decrease, as can be seen from the corresponding v_x line.

The y position looks roughly like a parabola, but a careful examination shows that it is no longer front-back symmetric, namely, the rate of rising before the top is different than rate of the falling after. The asymmetry would be also observable in the trajectory (y vs. x, not shown, Exercise 5.4). Of course, the range is shorter as well due to the x-position curve bending downward. Both velocity curves show non-constant slopes because of drag force changing with velocity. They level off with time, which, if allowed to continue sufficiently long, will reach a constant value, or terminal velocity. In addition, drag is a dissipative force,

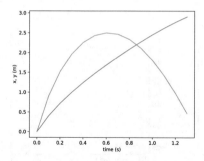 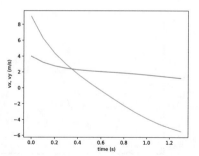

Fig. 5.2 Position- and velocity-time curves, respectively, in projectile motion with drag. Only quadratic drag is included, specifically, $c_1 = 0$ and $c_2 = 0.2$. The blue and orange colors correspond to $x-$ and $y-$ components, respectively

and mechanical energy will not be conserved as is in the ideal case. All of these effects will be enhanced if the drag coefficient is increased, showing the importance of drag relative to gravity. This will be explored in Exercise 5.4.

5.5   Periodic Motion

Having applied and tested the time-stepping method with projectile motion, we now can use the same technique to study a special kind of motion, periodic motion, which can be found all around us. Examples include the simple pendulum, planets orbiting the sun, heart beat, and so on. In this section we discuss two rather different yet closely related fundamental systems, the simple harmonic oscillator and planetary motion under gravity, the only two systems known to support closed orbits.

5.5.1   Simple Harmonic Oscillator

The simplest periodic motion is the simple harmonic oscillator (SHO), and arguably the most significant model in physics. The SHO consists of a particle of mass m attached to a spring with a spring constant k. The force is described by Hooke's law, $\mathbf{F} = -k\mathbf{r}$, where $\mathbf{r}$ is the displacement from equilibrium. This force is an empirical force as a result of electromagnetic interactions of many atoms, and not a fundamental force in nature.

We can write down the equations of motion for the SHO like before,

$$\frac{d\mathbf{r}}{dt} = \mathbf{v}, \quad \frac{d\mathbf{v}}{dt} = \mathbf{a}, \quad \mathbf{a} = -\omega^2 \mathbf{r}, \tag{5.14}$$

where $\omega = \sqrt{k/m}$ is the natural oscillation frequency. The solutions of the SHO are periodic and well-known, but it still benefits us to simulate it numerically.

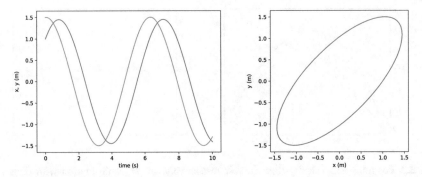

Fig. 5.3 Position-time curves (left) and the trajectory y vs. x (right) for a 2D simple harmonic oscillator. The blue and orange colors correspond to $x-$ and $y-$ components, respectively

Applying the Euler's method to Eq. (5.14) as before in similar cases, we can write an SHO program as shown in Program A.3.3 (also see GlowScript version in Appendix A). Overall, it is similar to other programs developed in this chapter. The output is shown in Fig. 5.3.

Figure 5.3 indeed shows sinusoidal oscillations in the position-time curves as expected. The frequency of oscillation is ω, and the period is $T = 2\pi/\omega$. In fact, it is helpful to rewrite the first-order ODEs of Eq. (5.14) as a compact second-order ODE,

$$\frac{d^2\mathbf{r}}{dt^2} = -\omega^2\mathbf{r}, \tag{5.15}$$

This is the well-known harmonic equation admitting sinusoidal solutions

$$r_i = A_i \sin(\omega t + \phi_i), \quad (i = x, y), \tag{5.16}$$

where A_i and ϕ_i are the amplitude and the phase, respectively, two constants for a second-order ODE to be determined from initial conditions. We leave it as an exercise to extract these parameters from Fig. 5.3 or Program A.3.3 (Exercise 5.5).

The trajectory is closed, meaning the particle will repeat the same orbit after each period. In fact, the orbit is an ellipse rotated about, and centered at, the origin. The shape, namely the eccentricity, and the orientation of the ellipse depend on the initial condition. Interestingly, the period is independent of the shape of the orbit or the initial conditions, in sharp contrast to planetary motion under gravity to be discussed next in Sect. 5.5.2.

The core of the program concerns this segment,

```
In [8]:  r = r + v*dt
         a = - omega**2 * r
         v = v + a*dt
```

where we update the position vector, then calculate acceleration using the latest position, and update velocity last. This is done intentionally.

It might seem that we should have calculated the acceleration first using the position value at the beginning of the step before the position update. However, if we had done so, the position would no longer be exactly sinusoidal, and the trajectory would not be a closed ellipse. Instead, the oscillation amplitudes would increase with time, and the trajectory would spiral outward. So the order is important. We leave it as an exercise (Exercise 5.5) for the reader to verify this.

The reason is because the Euler's method is not exact, and a small error can grow with each step, often growing exponentially in oscillatory motions. Fortunately, it can be shown that the error can be constrained by using the latest position to calculate acceleration and update velocity subsequently. Such a change of order in Euler's method is known as the Euler-Cromer method Eq. (4.10). The latter should be applied to physical systems whenever possible, but it is crucial to do so in periodic motion as well as in conservative systems.

That the Euler-Cromer method works so well can be qualitatively understood as follows. Imagine a particle connected to a spring is moving away from the equilibrium and slowing down. Suppose x_n is just one step before reaching the turning point (maximum stretch). The predicted next position x_{n+1} would be the turning point where the velocity should be zero. But if we used $a(x_n)$ to predict the next velocity, it would be close to but not be zero, causing an overshoot. Unphysical excess energy would build up. Instead, using $a(x_{n+1})$ would prevent the overshoot, and thus conserves energy.

Incidentally, the Euler-Cromer method is a member of the so-called symplectic methods. They preserve important physical properties including energy, closed path, and time reversal. These methods have high physics fidelity, and should be preferred over methods which may be of higher order but of less fidelity.

Similar to Eq. (5.6), we can express the Euler-Cromer method in vector form as follows,

$$\mathbf{r}_1 = \mathbf{r}_0 + \mathbf{v}_0 \times \Delta t,$$
$$\mathbf{v}_1 = \mathbf{v}_0 + \mathbf{a}(\mathbf{r}_1) \times \Delta t. \quad \text{(Euler-Cromer)} \quad (5.17)$$

This method is simple yet powerful enough for many problems as the next related example shows.

5.5.2 Planetary Motion

In astrophysical and planetary systems, motion is governed by Newton's law of gravity,

$$\mathbf{F} = -GMm\frac{\mathbf{r}}{r^3}. \quad (5.18)$$

Here, $\mathbf{F}$ is the gravitational force between two bodies of masses M and m, respectively, separated by a distance r, and G is the gravitational constant. The force is mutually attractive as indicated by the negative sign. If we take the position vector going from M to m, then Eq. (5.18) is the force acting on mass m. For simplicity, we assume m is much smaller than

M, i.e., $m \ll M$ in the following, such that M remains fixed in space, and m moves around M. Unlike the spring force in SHO, gravity is a fundamental force in nature.

With gravity given by Eq. (5.18), the equations of motion become

$$\frac{d\mathbf{r}}{dt} = \mathbf{v}, \quad \frac{d\mathbf{v}}{dt} = \mathbf{a}, \quad \mathbf{a} = -GM\frac{\mathbf{r}}{r^3}. \tag{5.19}$$

Note that the smaller mass m drops out after cancellation in acceleration $\mathbf{a} = \mathbf{F}/m$. Except for the specific form of acceleration, these equations are the same as those of SHO Eq. (5.14), so we expect the simulation programs to be also analogous.

But there is one important factor to consider before jumping in, scale. Astrophysical systems are large, so the use of SI units will be cumbersome and unnatural in numerical computation. Furthermore, these values in SI units are so astronomically large that the results often make little intuitive sense in terms of magnitude. It is much preferred to do the calculations in units natural to the system being considered.

To see this, let us consider a test body of mass m moving in a circular orbit of radius a around the larger mass M. The gravitational acceleration is equal to centripetal acceleration, and according to Eq. (5.19), we have $GMm/r^2 = mv^2/r$, or $GM = rv^2$. A natural unit for length is the radius a, and because the the motion is periodic, a natural unit for time is the period T. Furthermore, if the unit for speed is chosen to be $v_0 = a/T$, the speed of circular motion is $2\pi v_0$. Therefore, in these units, we have

$$GM = 4\pi^2, \quad \text{units: } av_0^2 = a^3/T^2. \tag{5.20}$$

Because GM always appears as one term, there is no need to define units of G and M separately.

For example, for the Earth-Sun system, we have $a = 1$ AU (astronomical unit), $T = 1$ year, and $GM = 4\pi^2$ AU^3/year. For convenience we may call these the solar unit system. Note that these natural units are scalable to other systems. For the Moon-Earth system, for instance, the unit of length would be the Earth-Moon separation, the unit of time approximately one month (27.3 d), and GM is still $4\pi^2$ in the scaled units, with M now referring to the mass of the Earth. We may call these the lunar unit system.

With the solar unit system, we can simulate the motion of any planet around the Sun via changing slightly Program A.3.3 to account for gravitational acceleration Eq. (5.19). The changes are minimal, the most essential part of the full Program A.3.4 is given below,

```
In [9]:  rmag = np.sqrt(r[0]**2 + r[1]**2)
         a = - GM * r/rmag**3
```

The above code just calculates acceleration from gravity instead of the spring force. Executing Program A.3.4 will produce results much more interesting than they first appear. For example, the program will display a closed elliptical orbit (not shown) similar to the orbit seen in Fig. 5.3 for the SHO. It is also remarkable that the forces from Hooke's law and gravity are the only two forces known to support closed orbits (see Goldstein et al. 2002).

However, there are several key differences. First, the center of gravity (Sun) is a focus of the ellipse instead of the center of the ellipse in the case of the SHO. Despite this, Kepler's second law, the radial vector sweeping out an equal area in equal time, still applies to both planetary motion and the SHO. This is because both forces are central forces so angular momentum is conserved. Second, the period of planetary motion is dependent on the shape and initial conditions rather than independent of either in the case of the SHO. Third, and the most significant, a planet can escape the gravitational field with sufficient energy, while SHO motion remains bound no matter how much energy it has. Gravity decreases as $1/r^2$ and when the energy is positive, the planet is deflected to infinity, also known as scattering. The famous Rutherford scattering led to the discovery of atomic nuclei because the Coulomb force between nuclei also decreases as $1/r^2$. Scattering is not possible in SHO because the Hooke's force increasing linearly with r is an empirical force stated earlier that does not exist as a fundamental force in nature, unlike gravity.

Program A.3.4 is generic and applicable to any gravitationally bound binary system, including the Earth-Sun or the Moon-Earth system (see the VPython version for animated planetary motion). We may wonder, then, how does one determine what planet is being simulated? The answer lies in the setting of initial conditions. For instance, if it is Earth we are interested in, we can set initial conditions in solar unit system as $\mathbf{r}_0 = [1, 0]$ and $\mathbf{v}_0 = [0, 2\pi]$, corresponding to placing Earth at the right side of the unit circle moving vertically up with Earth's orbital speed. Explore other initial conditions in Exercise 5.7.

5.5.3   Accuracy and Higher Order Methods

We have seen the Euler and Euler-Cromer methods yield reasonably accurate results. Accuracy is largely controlled by the step size Δt, but we have not discussed how we chose a particular value, except that it should be small. But small relative to what?

In every system, there is a typical (characteristic) time scale. For instance, a ball in free fall may last a few seconds, while it takes a year for the Earth moving around the Sun. We can estimate this characteristic time as $\tau_c \sim L_c/v_c$, where L_c is the characteristic size of the problem, and v_c the typical speed. To be accurate, we should choose $\Delta t/\tau_c \ll 1$ (say 0.01). But other factors should be considered, including highly oscillatory behavior or singularities, which may require even a smaller ratio.

Usually a smaller step size results in higher accuracy. But there is a trade-off between accuracy and speed. The number of steps to complete a simulation scales like $N \sim 1/\Delta t$, so the smaller the Δt, the longer the simulation. But reducing the step size is not the only way to increase accuracy. It is often possible to achieve the same goal with a higher-order method.

The Euler method, and its variant Euler-Cromer method, is a first-order method, meaning it is accurate to first order in Δt, and the local error is on the order of $O_1 \sim (\Delta t)^2$. There are schemes to achieving higher accuracy, including the well-known Runge-Kutta series (see Press et al. (2007)), with the Euler method being the first member.

Similarly, one can construct a sympletic series, with the Euler-Cromer method being the first member. The next member is the so-called leapfrog method (see Eq. (4.11)), which steps Eq. (5.7) forward in a leapfrogging fashion,

$$\mathbf{r}_m = \mathbf{r}_0 + \mathbf{v}_0 \times \frac{\Delta t}{2},$$
$$\mathbf{v}_1 = \mathbf{v}_0 + \mathbf{a}(\mathbf{r}_m) \times \Delta t,$$
$$\mathbf{r}_1 = \mathbf{r}_m + \mathbf{v}_1 \times \frac{\Delta t}{2}. \quad \text{(Leapfrog)} \tag{5.21}$$

The leapfrog method (also called the Verlet method) has three stages. First, half a step is taken to obtain the intermediate position $\mathbf{r}_m$ at the midpoint. Second, a full step is taken to find the new velocity v_1 using acceleration evaluated at the midpoint $\mathbf{a}(\mathbf{r}_m)$. Third, another half a step is taken to obtain the new position $\mathbf{r}_1$ using the new velocity. Each stage uses the newest values available from the previous stage.

The leapfrog method has one more stage than the Euler-Cromer method, but it is accurate to second order, namely, the local error is $O_2 \sim (\Delta t)^3$. The advantage of the leapfrog method over Euler-Cromer is significant. To see why, let $h_1 = (\Delta t)_1$ and $h_2 = (\Delta t)_2$ be the step sizes of the Euler-Cromer and leapfrog methods, respectively. If we want numerical solutions at a comparable error level, $O_1 \sim O_2$, we expect the step sizes to be related by $h_2^3 \propto h_1^2$, or $h_2 \sim \alpha h_1^{2/3}$, where α is a factor on the order of 1. The speed gain is given by $N_1/N_2 \sim h_2/h_1$. The following table shows the step size vs speed gain.

Table 5.1 shows that with each order of magnitude decrease in h_1, the speed gain increases by a factor of ~ 2.1. This is considerable for even moderately small h_1. The leapfrog method has just one more stage than the Euler-Cromer method, has the same fidelity, but offers significant speed gain. It is robust, accurate, and should be the method of choice for simulations of relatively simple systems (see Exercise 5.6).

Table 5.1 The step sizes and speed gain of leapfrog over the Euler-Cromer methods at comparable error levels

Euler-Cromer		Leapfrog		Speed gain
h_1	N_1	h_2	N_2	N_1/N_2
0.100	10.0	0.215	4.64	2.15
0.010	100	0.046	21.5	4.64
0.001	1000	0.010	100	10.0

5.6   Motion in Cross Electromagnetic Fields

Electromagnetic forces are another type of fundamental force, like gravity. However, the former is much stronger, so in motion driven by electromagnetic forces, we can usually ignore gravity. An important class of motion involves cross electromagnetic fields. We discuss this case next.

Electromagnetic fields exert two forces on a charged particle: an electric force due to the electric field $\mathbf{E}$, and the Lorentz force due to the magnetic field $\mathbf{B}$ (see Taylor 2005). The force and acceleration are

$$\mathbf{F} = q(\mathbf{E} + \mathbf{v} \times \mathbf{B}), \quad \mathbf{a} = \frac{q}{m}(\mathbf{E} + \mathbf{v} \times \mathbf{B}), \tag{5.22}$$

where q is the charge of the particle, m the mass, and $\mathbf{v}$ its velocity.

Note that the charge to mass ratio q/m appears as a single parameter in the acceleration. This ratio can be measured from motion in cross electromagnetic fields where the electric and magnetic fields are typically set up to be perpendicular to each other.

Unlike previous simulations discussed so far, the Lorentz force needs careful examination. Like drag Eq. (5.12), it depends on the velocity, but unlike drag, the Lorentz force is always perpendicular to the velocity because of the cross product. As a result, it can change the velocity of the particle but not its speed or kinetic energy because it does no work.

In numerical integration with a finite step, the change of the velocity due to the Lorentz force is $\Delta\mathbf{v} = \frac{q}{m}\mathbf{v} \times \mathbf{B}\Delta t$. Because $\Delta\mathbf{v}$ is perpendicular to $\mathbf{v}$, the resultant speed would be $|\mathbf{v}'| = \sqrt{|\mathbf{v}|^2 + |\Delta\mathbf{v}|^2} > |\mathbf{v}|$. Unless we take this problem into consideration, the energy would increase artificially, making the results inaccurate, or worse, causing instability.

We deal with the problem from two sides. First, as we have seen earlier Sect. 5.5.3, the leapfrog method is easy to use yet more accurate. We choose the leapfrog method over the first-order ones such as the Euler or Euler-Cromer methods, so overall higher accuracy can be obtained.

Second, we treat the electric force and the Lorentz force separately. The calculation due to the electric force can be done as usual. But in the calculation due to the Lorentz force, only the component $\mathbf{v}_\perp$ perpendicular to the magnetic field $\mathbf{B}$ is affected, again because of the cross product. Thus, we have $\mathbf{v}'_\perp = \mathbf{v}_\perp + \Delta\mathbf{v}$. We enforce speed to be constant by this scaling

$$\mathbf{v}'_\perp \frac{|\mathbf{v}_\perp|}{|\mathbf{v}'_\perp|} \Rightarrow \mathbf{v}'_\perp. \tag{5.23}$$

Now the new magnitude of the velocity is equal to the old magnitude $|\mathbf{v}'_\perp| = |\mathbf{v}_\perp|$, and has the correct direction. To obtain the perpendicular velocity, we first find the parallel velocity, $\mathbf{v}_\parallel = (\mathbf{v} \cdot \hat{\mathbf{B}})\hat{\mathbf{B}}$, then subtract it from $\mathbf{v}$, so $\mathbf{v}_\perp = \mathbf{v} - \mathbf{v}_\parallel$. See Program A.3.5 for actual implementation.

We implement the above strategy in Program A.3.5 (run the VPython version for animated motion). The results are shown in Fig. 5.4. The electric and magnetic fields are perpendicular

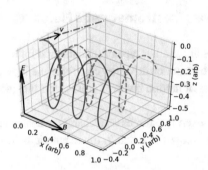

Fig. 5.4 Trajectories of a positively charged particle in cross electromagnetic fields. The electric and magnetic fields and the initial velocity are indicated by the respective arrows. The magnetic field and the initial velocity are the same in the three cases. The solid and dashed curves correspond to zero and intermediate electric fields, respectively. The dash-dot curve corresponds to an electric field that balances out the Lorentz force completely

to each other shown by the arrows. The particle enters the cross fields with a horizontal initial velocity ($v_z = 0$).

Results for three cases are displayed for three different electric fields. In the case of zero electric field, the only force is the Lorentz force (gravity is negligible) and motion is a typical spiral (helix), with circular motion in the $y - z$ plane and translational motion along the x direction. One can use this feature to measure the charge to mass ratio q/m if we know the radius of the circular motion.

When the electric field is increased to intermediate level, the spiral motion persists but it is no longer circular in the $y - z$ plane. Rather, it is elongated in the z direction like an ellipse. There is also a shift up and to the right due to the additional electric force. When the electric field is further increased so that it completely cancels the Lorentz force from the beginning, the motion is a straight line. This condition has many applications. For instance, it can be used as a velocity selector. Unless a particle entering the field has the right velocity, namely, $v_y = E/B$, it will be deflected up or down. Only particles with matching velocity will pass though unaffected.

Upon closer examination (see Exercise 5.9), we find that in the case of zero electric field, the circle projected onto the $y - z$ plane is closed and the speed remains constant to a high degree of accuracy. It shows that our strategy as implemented in Program A.3.5 works well. Compared to the standard leapfrog algorithm Eq. (5.21), the key modification occurs in the calculation of the velocity. Letting $\mathbf{a}_E = \frac{q}{m}\mathbf{E}$ and $\mathbf{a}_B = \frac{q}{m}\mathbf{v} \times \mathbf{B}$, the modified steps are,

$$\mathbf{v}_m = \mathbf{v}_0 + \mathbf{a}_E \times \frac{\Delta t}{2}$$
$$\mathbf{v}'_m = \mathbf{v}_m + \mathbf{a}_B \times \Delta t \tag{5.24}$$
$$\mathbf{v}_1 = \mathbf{v}'_m + \mathbf{a}_E \times \frac{\Delta t}{2}$$

We split the calculation of the electric field in two half-steps. We first take a half-step to calculate the midpoint velocity $\mathbf{v}_m$ due to the electric field alone. Next a full step is taken to obtain an updated midpoint velocity $\mathbf{v}'_m$ due to the magnetic field alone. Here a speed scaling Eq. (5.23) should also be applied so it remains constant after bending by the Lorentz force (see Program A.3.5). Finally we take another half-step with $\mathbf{a}_E$ to find the velocity $\mathbf{v}_1$ at the end. By splitting the calculation due to the electric field, it amounts to effectively using the average change of velocity due to $\mathbf{a}_E$ in the calculation with $\mathbf{a}_B$. This results in a higher accuracy. Program A.3.5 can be a template enabling further explorations of motion in electromagnetic fields, including validation of energy conservation (Exercise 5.9), and motion in nonuniform, even time-dependent fields.

5.7 Exercises

5.1. **Kinematic parameters from fitting.** Fitting motion data to obtain kinematic parameters such as velocity and acceleration is an effective, and sometimes necessary, method. We explorer data fitting in this exercise.

(a) In Sect. 5.2, motion with constant acceleration is modeled in Program 5.2 with Euler's method and the results are shown in Fig. 5.1. Choose your own initial values in Program 5.2. For instance, set acceleration to be negative, predict what the position- and velocity-time curves would look like, and sketch them. Run the program to check against your predictions.

(b) Now that we are familiar with Program 5.2, modify it to fit the position-time data to a quadratic relation to extract initial position, velocity, and acceleration using the curve-fitting technique discussed in Sect. 4.5. Compare the best-fit values with the initial values assumed in Program 5.2. Overlay the analytic results Eq. (5.8). Discuss the accuracy of the simulation (Euler's method) in terms of the comparison.

(c) Fit the velocity-time data to a linear relation to obtain the initial velocity and acceleration. Are these values consistent with the position-time fit? Explain any difference.

5.2. **Validating projectile motion.** Graph the analytic position- and velocity-time results for ideal projectile motion along side the numerical ones in Program A.3.1.

Optionally, add a statement inside the loop to calculate and record the mechanical energy (kinetic plus potential, assuming mass $m = 1$ kg). Plot energy versus time. Is it conserved as expected?

5.3. Swoosh that shot. Determine graphically the angle required to make a basketball shot. Assume launching speed $v_0 = 7$ m/s, and try to hit the spot at $R = 3$ m away and $h = 1.5$ m above the release point. Neglecting drag, the trajectory of the ball is

$$y(x) = \tan \theta \, x - \frac{1}{2} g \frac{x^2}{v_0^2 \cos^2 \theta}.$$

The main plotting function should accept angle θ as the input parameter. A possible pseudo code for it is

```
def ploty(theta):
    convert theta to radians
    compute y on the x-grid
    plot y vs x
    mark the height h (at x=R) with an arrow
    set ylim to [0, 2] for easy visual

initialize v0, g set up x grid over [0, R]
interactive(ploty, ...)
```

Use the `interactive` function with IPywidget (Sect. 3.1.2). Drag the slider for θ to find the angles where $y(R) = h$. The angles should be accurate to within $0.2°$.

Note that only at discrete angles can the trajectory bend correctly so as to hit the right spot. Does the angle being discrete surprise you? Explain why.

Vary other parameters such as the initial speed or the target spot, and repeat. One could call this the "quantization" of the angle. We will again see something similar in quantum mechanics in the form of energy quantization (Sect. 7.3).

5.4. Effects of air resistance. We expand on the investigation of drag effects in realistic projectile motion (Sect. 5.4).

(a) Using existing parameters, plot the y vs. x curve with Program A.3.2. Explain the forward-backward asymmetry. Of the horizontal and vertical drag forces, is one more important than the other in terms of inducing the asymmetry? How can we test it?

(b) Calculate and graph the energy E (kinetic plus potential) as a function of time (safely assuming unit mass m, or just think of it as E/m, energy per unit mass).

*(c) The loss of energy is due to work done by drag, W_d. The cumulative work W_d can be calculated in the loop as (again work per unit mass) $W_d/m = \sum_i w_i$, where

$w_i = \mathbf{a}_d \cdot \Delta\mathbf{r}$ is the work done in the time interval Δt. Both $\mathbf{a}_d$ and $\mathbf{r}$ are available in the loop. The pseudo code looks like this,

```
ad = − c1*v − c2*speed*v
dW = np.dot(ad, dr)
add dW to W and append W
```

Compute W_d and plot it together with E above. Also plot the difference $E − W_d$. Explain the result.

*(d) In Program A.3.2, we assumed a c_2 value without justification. It is given by roughly $c_2 \sim \rho A/m$, with ρ being the density of air ($1.2\,\text{kg/m}^3$) and A the cross sectional area of the object. Estimate the c_2 value for a soccer ball of radius $0.11\,\text{m}$ and mass $0.4\,\text{kg}$. Find the range and height of a soccer ball kicked with $30\,\text{m/s}$ off the ground at $40°$ angle.

*(e) Imagine a soccer ball is launched at the top of a high cliff and falls into a deep chasm. It should eventually reach terminal velocity when drag and gravity balance each other. Modify Program A.3.2 to find the terminal velocity.

Integrate the motion until the speed reaches a plateau. Numerically, if the change of speed Δv between two steps is smaller than certain tolerance ϵ, $\Delta v/v \le \epsilon$, we can say it has plateaued. Test convergence with $\epsilon = 10^{-3}$ or 10^{-4}.

5.5. SHO orbits.

(a) Instead of using the Euler-Cromer method in Program A.3.3 for the simple harmonic oscillator (SHO), change it to the Euler method. To do so, just switch the order of the position and acceleration statements in the main loop so acceleration is calculated first. Leave the rest unchanged. Run the program and graph the results similar to Fig. 5.3. Is the orbit closed? Are the results physically correct? Explain (also see Exercise 4.3).

(b) Fit the positions x and y to obtain the amplitudes $A_{x,y}$ and phases $\phi_{x,y}$ according to Eq. (5.16). Optionally graph the fitted curves with the actual ones on the same plot. What initial condition would lead to the same amplitudes $A_x = A_y$? Test it. Is it also possible to have the same phases $\phi_x = \phi_y$?

5.6. Leapfrog integrator. Implement the leapfrog method Eq. (5.21) in the SHO simulation, Program A.3.3. Check that it works correctly and produces the same results as shown in Fig. 5.3.

Calculate the energy E in the main loop and plot the difference $E − E_0$, with E_0 being the initial energy, as a function of time with both the Euler-Cromer and the leapfrog methods. Recall the energy per unit mass of an SHO is $E/m = \frac{1}{2}(\omega^2 r^2 + v^2)$. A suggested pseudo code is,

```
def energy(r, v):
    r2 = np.dot(r, r)   # r squared
    v2 = np.dot(v, v)   # v squared
    return 0.5 * (omega**2 * r2 + v2)
......
E0 = energy(r, v)
while t< 10.0:
    ......
    E = energy (r, v)
    append to E list
    ......

convert E list to Earray = np.array(E list)
plot Earray−E0 vs time
```

Comparing the small oscillations, comment on the accuracy of the two methods.

5.7. Precession of planets. Explore planet motion with Program A.3.4. The default initial conditions lead to a nearly circular orbit. Now, change the initial position to $\mathbf{r} = [1.0, 0.5]$, run the simulation. Observe the orbit shape. Next, change it to $\mathbf{r} = [0.6, 0.0]$ and repeat. You should see orbits rotating around, called precession. While orbital precession can be real in non-two-body motion, the apparent precession here is due to low-order inaccuracy of the integrator, that is to say, it is a numerical artifact. Optionally, investigate this artificial precession further by decreasing the step size Δt, or using a more accurate integrator (Exercise 5.8).

5.8. Planet motion.

(a) Time the main loop in the planet simulation Program A.3.4 (see Program A.1.8). Increase the final time (e.g. to t < 10000.0) such that it takes about 10 s to finish using the Euler-Cromer method. Also note the ring-like spread (width) of the trajectory "belt".
Next, implement the leapfrog method in Program A.3.4. Time the main loop again, and compare the results. Especially comment on the width of the belt.

(b) Calculate the energy, and optionally the angular momentum, in planet motion with both methods discussed above. Plot them as a function of time. Discuss the results in terms of accuracy.

*(c) You should have noticed that the leapfrog method took about 50% more time in part 5.7 but was much higher in accuracy. According to Table 5.1, for the *same* accuracy, leapfrog should be faster than the Euler-Cromer method.
Use the corresponding step sizes listed in the table for the same expected accuracy, run the program again. Reduce the step size to 0.001 and repeat. You should record

and plot fewer points to reduce latency, say 1 in 1000 points when Δt is this small. Describe the results of both methods in terms of speed and accuracy.

5.9. Circular motion in magnetic fields. Motion of a charged particle is circular in a pure magnetic field (Fig. 5.4). We will investigate the motion in some detail next.

(a) Run Program A.3.5 with the same parameters except setting electric field to zero. The only force acting on the particle is the Lorentz force. Is the energy of the particle conserved?

Add a statement in the main loop of Program A.3.5 to calculate and record the kinetic energy K_i as a function of time. Assume unit mass. Graph the kinetic energy, and the relative error with respect to initial kinetic energy, $|K_i - K_0|/K_0$. Comment on the result. How well is energy conserved?

(b) Again run Program A.3.5 with zero electric field. Project the trajectory onto the $y - z$ plane by plotting z vs. y only. You should also set equal aspect ratio (see Program A.3.4). Does it appear circular? Is the path closed? Use the interactive zoom tool to see finer details. Describe your observations.

Read off the graph the center of the circle and the radius. Write them down.

(c) Think about the effect of speed and the magnetic field strength on the radius. Make a prediction on how much the radius will change if any of the following is doubled: (i) the magnetic field alone, (ii) the speed alone, and (iii) both simultaneously. Discuss with your partner or instructor if available. Write down your predictions.

Repeat part 5.9b, test your predictions, and describe your findings.

*(d) Obtain accurately the circle parameters. There are three parameters determining a circle $(y - a)^2 + (z - b)^2 - r^2 = 0$, the center being (a, b) and the radius r. We can pick three arbitrary points (y_i, z_i) to uniquely determine these parameters. From the position data after the main loop in Program A.3.5, choose three points some steps apart, say $s = 50$ steps from the start. Set up the equations as follows,

$$\text{eqns} = [(y_i - a)^2 + (z_i - b)^2 - r^2 = 0, \quad i = 0, s, 2s]$$

Solve eqns with Sympy as (Sect. 3.3)

```
In [10]: eqns = [(y[i*s]-a)**2 + (z[i*s]-b)**2 - r*r for i in range(0,3)]
         solve(eqns, a, b, r)
```

Compare the numerical values a, b, r with the graphical readout of part 5.9b. Optionally, vary the three-point set (y_i, z_i) to check the variance in the values, and compare with theoretically expected result for the radius, $r = m|\mathbf{v}_\perp|/qB$.

References

C. Frohlich. Aerodynamic drag crisis and its possible effect on the flight of baseballs. *Am. J. Phys.*, **52**: 325–334, (1984).

H. Goldstein, C. Poole, and J. Safko. *Classical mechanics*. (Addison Wesley, New York), 2002.

W. H. Press, S. A. Teukolsky, W. T. Vetterling, and B. P. Flannery. *Numerical recipes: the art of scientific computing*. (Cambridge University Press, Cambridge), 3rd edition, 2007.

J. R. Taylor. *Classical mechanics*. (University Science Books, New York), 2005.

Oscillations and Waves

Oscillations and waves are not only common in nature but they are also inseparable. When one plucks a string, atoms in the immediate vicinity will start to jiggle around, pushing or tugging neighboring atoms into similar jiggling motion. This creates a wave propagating along the string. In essence, a mechanical wave is a result of collective oscillations of many particles carrying energy as it propagates.

In Chap. 5 we discussed the periodic oscillation of a single simple harmonic oscillator. Now we turn our attention first to the oscillations of few-body systems, also known as coupled oscillators. We will study characteristics of coupled oscillators including normal modes and frequencies. Subsequently we will discuss simulations of waves in a string where we will see some of the same characteristics.

6.1 Coupled Oscillators

We have seen the simulation of a single simple harmonic oscillator (SHO) in Sect. 5.5.1 and now seek to expand on it to include one or more SHOs coupled to each other.

6.1.1 Two-Body Oscillations

Let us first consider a two-body system consisting of particles of respective masses m_1 and m_2 connected by a massless spring with spring constant k. For simplicity, we restrict the motion to one dimension as illustrated in Fig. 6.1a. Unlike the 2D motion considered earlier in Sect. 5.5.1, the spring is now connected to two moving particles instead of one end fixed to the origin. Now both particles are coupled to each other as coupled oscillators.

© The Author(s), under exclusive license to Springer Nature Switzerland AG 2023   115
J. Wang and A. Wang, *Introduction to Computation in Physical Sciences*, Synthesis Lectures
on Computation and Analytics,
https://doi.org/10.1007/978-3-031-17646-3_6

Fig. 6.1 Coupled oscillators. (a) a two-body system; and (b) a three-body system

What are the equations of motion for the coupled oscillators? As seen from Eq. (5.14) for the SHO, we need two pairs of position and velocity variables. For position, we could use the absolute positions of the particles from the origin. Instead, it is actually more convenient to use the relative positions of the particle from equilibrium because, to calculate the spring force, we now would not need the length of the unstretched spring to figure out the amount of stretching or compression of the spring, one fewer nonessential parameter to carry around.

Let u_1 and u_2 denote the positions (displacements) of m_1 and m_2 from their equilibrium points, respectively. From Fig. 6.1a, the difference, $u_2 - u_1$, measures the net stretching (or compression if it is negative) of the spring. The force on m_1 from m_2 is $F_{12} = k(u_2 - u_1)$, and from Newton's third law, the force on m_2 from m_1 is $F_{21} = -F_{12} = k(u_1 - u_2)$. Though the forces now involve two variables, they are still restoring forces like the standard spring force on a single particle. Take F_{21} for example, for a given u_1, it grows more negative as u_2 becomes more positive, pulling m_2 toward equilibrium harder, and vice versa.

We have enough information to start simulating the two-body oscillator. However, we take a different approach here and look at a few analytic properties first because it is instructive going forward. Here is an important example of making necessary connections between analytical and computational approaches to fully understand the models and results.

Let us start with the second-order harmonic equation similar to Eq. (5.15),

$$\frac{d^2 u_1}{dt^2} = -\frac{k}{m_1}(u_1 - u_2), \quad \frac{d^2 u_2}{dt^2} = -\frac{k}{m_2}(u_2 - u_1). \tag{6.1}$$

The right-hand sides are just accelerations from the respective forces.

Given the restoring nature of the forces, it is reasonable to expect harmonic oscillations in the two-body system as well. So let us use Eq. (5.16) as trial solutions, $u_i = A_i \sin(\omega t + \phi_i)$, $i = 1, 2$, with ω to be determined. Using $d^2 u_i / dt^2 = -\lambda u_i$ where $\lambda = \omega^2$, substituting the trial solutions into Eq. (6.1), and after some rearranging, we obtain,

$$\frac{k}{m_1} u_1 - \frac{k}{m_1} u_2 = \lambda u_1, \tag{6.2a}$$

$$-\frac{k}{m_2} u_1 + \frac{k}{m_2} u_2 = \lambda u_2. \tag{6.2b}$$

Here we have two linear equations to solve. Suppose we set out this way: express u_2 in terms of u_1 from Eq. (6.2a), and substitute it into Eq. (6.2b) to obtain u_1, we have

$$\left[\left(\frac{k}{m_1} - \lambda\right)\left(\frac{k}{m_2} - \lambda\right) - \frac{k^2}{m_1 m_2}\right] \times u_1 = 0. \tag{6.3}$$

This relationship is interesting in that we sought to obtain u_1 but ended up with a product of two terms equal to zero, so either the term in $[\cdots] = 0$ or $u_1 = 0$. The latter is a null solution, not physically acceptable. The former would place restrictions on λ (or ω) that only select, discrete values are possible. Evidently, the relationship Eq. (6.3) and these discrete values are in fact well-known mathematically as the eigenequation (or characteristic equation) and eigenvalues, respectively.

Solving the quadratic equation $[\cdots] = 0$ in Eq. (6.3) yields two eigenvalues,

$$\lambda_{1,2} = \left[0, \frac{k}{\mu_{12}}\right], \quad \mu_{12} = \frac{m_1 m_2}{m1 + m2}. \tag{6.4}$$

We recognize that μ_{12} is the reduced mass between m_1 and m_2. The characteristic frequencies $\omega = \sqrt{\lambda} = [0, \sqrt{k/\mu_{12}}]$ are called eigenfrequencies (or normal frequencies).

Before discussing the physical meaning of all this, let us take a detour briefly reviewing the eigenvalue problem from the perspective of linear algebra, of which Eqs. (6.2a) and (6.2b) are but one illustrative example. It is standard to rewrite those equations in matrix form,

$$\mathbf{Au} = \lambda\mathbf{u}, \quad \mathbf{A} = \begin{bmatrix} \frac{k}{m_1} & -\frac{k}{m_1} \\ -\frac{k}{m_2} & \frac{k}{m_2} \end{bmatrix}, \quad \mathbf{u} = \begin{bmatrix} u_1 \\ u_2 \end{bmatrix}. \tag{6.5}$$

The matrix $\mathbf{A}$ is square, and called the system matrix (or stiffness matrix). The column matrix $\mathbf{u}$ containing the solutions is called the state vector or eigenvector. Equation (6.5) is a special linear system because the effect of $\mathbf{A}$ acting on $\mathbf{u}$ must result in the same state vector, up to an overall scaling factor λ. Apparently, not any but only select values of λ can satisfy the requirement, namely,

$$\det|\mathbf{A} - \lambda\mathbf{1}| = 0. \tag{6.6}$$

In principle the determinant of Eq. (6.6) produces a polynomial of degree N equal to the dimension of $\mathbf{A}$, resulting in N roots, of which Eq. (6.3) is a special example of $N = 2$. When N is small, such as the present case Eq. (6.4), solving the polynomial equation directly is an option. But it is impractical for slightly larger N. Well-developed computational schemes are ideally suited for this task as we will see next.

Once the eigenvalues are obtained, regardless of how, we can substitute them into Eq. (6.5) to find the eigenvector corresponding to a given eigenvalue, up to an overall constant. This constant may optionally be determined by a procedure known as normalization. In our case, plugging the eigenvalues Eq. (6.4) into Eqs. (6.2a) and (6.2b) give us these results,

$$\lambda_1 = 0 : \mathbf{u}_1 = \begin{bmatrix} 1 \\ 1 \end{bmatrix}; \quad \lambda_2 = \frac{k}{\mu_{12}} : \mathbf{u}_2 = \begin{bmatrix} -\frac{m_2}{m_1} \\ 1 \end{bmatrix}. \tag{6.7}$$

6.1.2 Normal Modes

Together, the eigenfrequencies and eigenvectors Eq. (6.7) are called normal modes (or eigen-modes), and they are the fundamental modes (see Fowles and Cassiday (2004)). What kind of motion do these normal modes represent? The eigenvectors give the relative proportion of position u_1 and u_2. In the first normal mode of Eq. (6.7), $\omega_1 = 0$, the motion is not oscillatory at all. Because $u_1 = u_2$ at all times, this mode represents two particles moving uniformly in translational motion, either at rest or at constant velocity, i.e.,

$$\omega_1 = 0: \quad \mathbf{u}_1(t) = \begin{bmatrix} 1 \\ 1 \end{bmatrix} \times (u_{cm} + v_{cm}t). \tag{6.8}$$

We can ascribe this to the motion of the center of mass because there is no net external force. As seen earlier from Sect. 5.1, the displacements will be linear if acceleration is zero. Consequently, the two constants u_{cm} and v_{cm} represent the displacement and velocity of the center of mass, respectively.

In the second normal mode of Eq. (6.7), $\omega_2 \neq 0$, the full solution must include the sinusoidal time dependence assumed earlier,

$$\omega_2 = \sqrt{\frac{k}{\mu_{12}}}: \quad \mathbf{u}_2(t) = \begin{bmatrix} -\frac{m_2}{m_1} \\ 1 \end{bmatrix} \times A \sin(\omega_2 t + \phi). \tag{6.9}$$

This mode represents two particles oscillating in opposite directions, proportioned by $u_1/u_2 = -m_2/m_1$. The negative sign indicates that u_1 and u_2 are $180°$ out of phase with each other, on the opposite sides of the center of mass. The same is also true for the velocities $\dot{u}_1$ and $\dot{u}_2$. We ascribe this to the oscillation about the center of mass, because $m_1 u_1 + m_2 u_2 = 0$ at all times. The constants A and ϕ are the amplitude and phase, respectively.

These two normal modes can completely describe the motion of the two-body system. In other words, any motion may be broken down into a linear combination of these two fundamental modes,

$$\mathbf{u}(t) = \mathbf{u}_1(t) + \mathbf{u}_2(t) = \begin{bmatrix} u_1(t) \\ u_2(t) \end{bmatrix}, \tag{6.10}$$

with constants in Eqs. (6.8) and (6.9) determined from initial conditions, $u_1(0)$, $u_2(0)$, $\dot{u}_1(0)$ and $\dot{u}_2(0)$ as follows (Exercise 6.1):

$$u_{cm} = \frac{m_1 u_1(0) + m_2 u_2(0)}{m_1 + m_2}, \quad v_{cm} = \frac{m_1 \dot{u}_1(0) + m_2 \dot{u}_2(0)}{m_1 + m_2},$$

$$\phi = \arctan\left(\omega_2 \frac{u_1(0) - u_2(0)}{\dot{u}_1(0) - \dot{u}_2(0)}\right), \quad A = \frac{m_1[u_2(0) - u_1(0)]}{(m_1 + m_2)\sin\phi}. \tag{6.11}$$

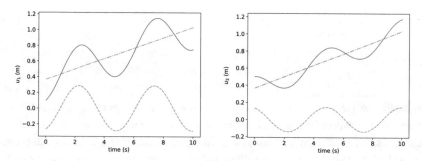

Fig. 6.2 Displacement versus time of a two-body coupled oscillator for u_1 (left) and u_2 (right). In each case, the solid curve is the net displacement, the dash-dot curve the normal mode $\mathbf{u}_1$ (center of mass motion), and the dashed curve the normal mode $\mathbf{u}_2$ (oscillation about center of mass). The parameters used are $m_1 = 1$ kg, $m_2 = 2$ kg, and $k = 1$ N/m. The initial conditions are $u_1(0) = 0.1$ m, $u_2(0) = 0.5$ m, $\dot{u}_1(0) = 0.2$ m/s, $\dot{u}_2(0) = 0$

To illustrate these normal modes, we show simulation results in Fig. 6.2 obtained from Program A.4.2 (see Exercise 6.2). The displacements u_1 and u_2 of each particle are shown in separate figures. In each figure, the net displacement shows an oscillatory component on top of a steadily increasing trend line. Indeed, the trend line is the first normal mode $\mathbf{u}_1(t)$ due to the center of mass motion Eq. (6.8), and the oscillation is the second normal mode $\mathbf{u}_2(t)$ Eq. (6.9). When the two normal modes as determined from the initial conditions Eq. (6.11) are added together, they reproduce the net displacement.

The results and analysis illustrate a general rule that any two-body problem can be reduced to an effective one-body problem in the center of mass frame. Therefore, true many-body effects must involve at least three bodies.

Symbolic Computation

Having discussed the two-body system in some detail, we are now in position to extend it to three or more bodies. Before doing so, this is a good time to review Sect. 3.3 on symbolic computation with Sympy because, as stated earlier, difficulties in the eigenvalue problem will increase significantly when N is even moderately larger, and manual calculations become intractable quickly.

Numerical computation can be enhanced significantly with symbolic computation in general, and with Sympy in particular, a versatile library for symbolic algebra in Python (Sect. 3.3). Here we demonstrate its application in the eigenvalue problem but its scope and capability extends way beyond this (see Chap. 7). The following codes are from Program A.4.1.

```
In [1]:  from sympy import *        # import symbolic lib sympy
         init_printing()            # pretty output
```

```
In [2]:  # declare variables
         m1, m2, m3, m, M = symbols("m_1 m_2 m_3 m M", real=True, positive=True)
         k, k1, k2 = symbols("k k1 k2", real=True, positive=True)
```

In the first cell above, we import sympy and set up "pretty" printing so output can show common math symbols including Greek letters and sub- or super-scripts. Next, we declare variables with symbols so they can be used in algebraic expressions. Additional properties associated with a variable can be specified with parameters such as real and positive attributes shown.

Below we construct the system matrix A from Eq. (6.5) and simply invoke the built-in eigenvals and eigenvects methods to obtain the eigenvalues and eigenvectors.

```
In [3]:  A = Matrix([[k/m1,-k/m1],[-k/m2,k/m2]])        # 2 x 2 system matrix
         A
```

Out[3]: $\begin{bmatrix} \dfrac{k}{m_1} & -\dfrac{k}{m_1} \\ -\dfrac{k}{m_2} & \dfrac{k}{m_2} \end{bmatrix}$

```
In [4]:  vals = A.eigenvals()       # eigenvalues only
         vals
```

Out[4]: $\left\{ 0 : 1, \quad \dfrac{k(m_1+m_2)}{m_1 m_2} : 1 \right\}$

```
In [5]:  vecs = A.eigenvects()      # eigenvectors
         vecs
```

Out[5]: $\left[\left(0, \quad 1, \quad \left[\begin{bmatrix} 1 \\ 1 \end{bmatrix} \right] \right), \quad \left(\dfrac{k(m_1+m_2)}{m_1 m_2}, \quad 1, \quad \left[\begin{bmatrix} \dfrac{m_2}{k}\left(\dfrac{k}{m_2} - \dfrac{k(m_1+m_2)}{m_1 m_2} \right) \\ 1 \end{bmatrix} \right] \right) \right]$

The eigenvalue output cell shows two-term groups, one group for each eigenvalue. The first term in a group is the eigenvalue, and the second term the multiplicity, namely, the number of times the same eigenvalue occurs. In our case, each eigenvalue occurs only once, hence the multiplicity is 1. It can happen that the multiplicity is greater than one, then the system is said to be degenerate. The two-body system is non-degenerate. The two eigenvalues agree with our manual calculation Eq. (6.4).

The eigenvector output is given in three-term groups: eigenvalue, multiplicity, and eigenvector. The first eigenvector belonging to the eigenvalue zero is exactly as before Eq. (6.8), but the second one is not in simplified form, though it should be the same as in Eq. (6.9).

Despite the power of symbolic capability, sympy often needs a helping hand simplifying the results, sometimes manually. There are built-in methods for simplification including simplify, factor, and combsimp, but the degree of success depends on the expression being simplified. In our case, factorization seems to give the best result because the result is reduced to the simplest form of Eq. (6.9).

```
In [6]: factor(vecs)      # simplify results, also try simplify(), combsimp()
```

$$\text{Out[6]:} \quad \left[\left(0, \quad 1, \quad \left[\begin{bmatrix} 1 \\ 1 \end{bmatrix}\right]\right), \quad \left(\frac{k(m_1+m_2)}{m_1 m_2}, \quad 1, \quad \left[\begin{bmatrix} -\frac{m_2}{m_1} \\ 1 \end{bmatrix}\right]\right)\right]$$

The normal modes analysis helps us understand the fundamental motion inherent in the two-body simulation results (Fig. 6.2). Symbolic computation for the two-body system may be a luxury, but for N-body systems including $N \geq 3$ it is indispensable. We examine three-body oscillators next.

6.1.3   Three-Body Oscillations

As stated earlier, two-body systems are effectively one-body problems with reduced mass. Nontrivial many-body effects would require at least three bodies, which we consider in this section. We will first examine the normal modes with the aid of symbolic calculation. We then simulate the three-body motion and analyze the results in terms of the normal modes.

Normal Modes

Adding a third mass m_3 and the corresponding u_3 to the two-body system (Fig. 6.1b), we now have a three-body system with two springs. If we assume the spring constants to be k_1 and k_2, there will be another third-law pair of forces between m_2 and m_3, $F_{23} = k_2(u_3 - u_2)$, in addition to $F_{12} = k_1(u_2 - u_1)$ between m_1 and m_2.

Knowing the respective accelerations from the forces, we can generalize the equations of motion from Eq. (6.1) and set up the eigenvalue problem similar to Eqs. (6.2a), (6.2b), and (6.5). Solving it with sympy, and after manual simplification, the eigenvalues are (see Exercise 6.3),

$$\lambda = \left[0, \; \frac{1}{2}\left(\frac{k_1}{\mu_{12}} + \frac{k_2}{\mu_{23}} \mp \sqrt{\left(\frac{k_1}{\mu_{12}} - \frac{k_2}{\mu_{23}}\right)^2 + \frac{4k_1 k_2}{m_2^2}}\right)\right]; \quad \mu_{ij} = \frac{m_i m_j}{m_i + m_j} \quad (6.12)$$

There are three eigenvalues as expected, including the value of zero. Again, the zero eigenvalue indicates uniform translational motion of the center of mass. The other nonzero ones involve collective three-body motion that cannot be reduced to two-body motion in general, showing coupling of all three bodies and true many-body effects.

If we restrict the parameters in Eq. (6.12), we can obtain interesting special cases. Some of these cases are listed here.

λ	Condition	
$\left[0, \dfrac{k_1}{\mu_{12}}\right]$	$k_2 = 0$	(6.13a)
$\left[0, \dfrac{k_{12}}{\mu_{13}}\right]$	$m_2 = 0, \ k_{12} = \dfrac{k_1 k_2}{k_1 + k_2}$	(6.13b)
$\left[0, \dfrac{k_1}{m_1}, \dfrac{k_2}{m_3}\right]$	$m_2 = \infty$	(6.13c)
$\left[0, \dfrac{k}{m}, \dfrac{3k}{m}\right]$	$k_1 = k_2 = k, \ m_1 = m_2 = m_3 = m$	(6.13d)
$\left[0, \dfrac{k}{m}, \dfrac{\tilde{k}}{\tilde{\mu}}\right]$	$k_1 = k_2 = k, \ m_1 = m_3 = m, \ m_2 = M, \ \tilde{k} = 2k, \ \tilde{\mu} = \dfrac{2m \times M}{2m + M}$	(6.13e)

In Eq. (6.13a), the spring between m_2 and m_3 is severed because $k_2 = 0$, so the results are the same as that of the two-body system Eq. (6.4). Similarly, in the limit $m_2 \to 0$ Eq. (6.13b), the middle mass is removed, we again obtain a two-body system with analogous results.

Note, however, for identical springs with $k_1 = k_2 = k$ in Eq. (6.13b), the effective spring length between m_1 and m_3 would be doubled, and the effective spring constant would be halved $k_{12} = k/2$. Two springs in series would make a longer spring that is easier to stretch and thus weaker. Conversely, cutting a spring in half would make it stronger, doubling the spring constant. Therefore, if we cut a piece of length L from a spool of uniform spring, we expect $kL = T$, with k being the spring constant of the piece, and T some constant having the dimensions of force and dependent on the type of material. (The constant turns out to be Young's modulus. This point is relevant to the wave equation Eq. (6.18) to be discussed in Sect. 6.2.)

In the opposite limit $m_2 \to \infty$ Eq. (6.13c), the middle mass does not move, and m_1 and m_3 behave like two single, uncoupled SHOs, discussed in Section 5.5.1. Finally, Eq. (6.13d) represents identical springs and masses, and Eq. (6.13e) identical springs but a different middle mass. Incidentally, the latter is a good model for studying vibrations of symmetric triatomic molecules like CO_2.

In general, except the center of mass motion (zero frequency), we need to analyze the eigenvectors to figure out the oscillations represented by each normal mode. For example, the normal modes corresponding to special case Eq. (6.13e) as obtained from Program A.4.1 are,

$$\lambda_1 = 0: \ \mathbf{u}_1 = \begin{bmatrix} 1 \\ 1 \\ 1 \end{bmatrix}; \quad \lambda_2 = \frac{k}{m}: \ \mathbf{u}_2 = \begin{bmatrix} -1 \\ 0 \\ 1 \end{bmatrix}; \quad \lambda_3 = \frac{\tilde{k}}{\tilde{\mu}}: \ \mathbf{u}_3 = \begin{bmatrix} 1 \\ -\frac{2m}{M} \\ 1 \end{bmatrix}. \quad (6.14)$$

The normal mode $\mathbf{u}_2$ shows that the outer masses stretches symmetrically in opposite directions while the middle mass stays at rest. In the third normal mode $\mathbf{u}_3$, the outer masses move in the same direction and the middle mass move in the opposite direction. One spring is being compressed and the other being stretched simultaneously. We note that these normal modes of classical oscillation correspond to vibrational states of molecules such as CO_2 in a quantum mechanical description (Sect. 7.3.3), and are the cause of greenhouse effects via excitation of the vibrational states (see Sect. 8.5). To see the normal modes dynamically, we discuss numerical modeling next.

Numerical simulation

Following Eq. (6.1), we need to add the third mass and convert the second-order equations to first-order ones to facilitate numerical integration. Denoting the velocities of $m_{1,2,3}$ by $v_{1,2,3}$, respectively, the resulting equations of motion for the three-body system (Fig. 6.1b) are,

$$
\begin{aligned}
\frac{du_1}{dt} &= v_1, & \frac{dv_1}{dt} &= -\frac{k_1}{m_1}(u_1 - u_2), \\
\frac{du_2}{dt} &= v_2, & \frac{dv_2}{dt} &= -\frac{k_1}{m_2}(u_2 - u_1) - \frac{k_2}{m_2}(u_2 - u_3), \\
\frac{du_3}{dt} &= v_3, & \frac{dv_3}{dt} &= -\frac{k}{m_3}(u_3 - u_2).
\end{aligned}
\tag{6.15}
$$

Compared to Eq. (5.14), the number of equations is tripled and the variables are no longer vectors. The acceleration of the middle mass has two terms due to the additional force F_{23}. But significantly, the variables are coupled, namely, they are interdependent because one particle affects the motion of the other in a nontrivial manner.

However, if we treat the three variables as components of pseudo vectors, namely, $\mathbf{u} = [u_1, u_2, u_3]$ and similarly for $\mathbf{v}$ and $\mathbf{a}$, we can just copy the template of Program A.3.3 to obtain the program for the three-body oscillator. As shown in the full Program A.4.2, the two programs are nearly identical in structure, with one significant difference in the time-stepping method, the leapfrog method:

```
In [7]:  u = u + 0.5*v*dt          # leapfrog
         f12 = -k1*(u[0]-u[1])
         f23 = -k2*(u[1]-u[2])
         a = np.array([f12/m1, -f12/m2 + f23/m2, -f23/m3])
         v = v + a*dt
         u = u + 0.5*v*dt
```

To deal with potentially highly oscillatory behavior of the three-body system, we use the leapfrog method Eq. (5.21) because it is more accurate than Euler-Cromer but requires only one more line of code. The displacement u is first evaluated over half a step; forces and acceleration calculated at the latest displacement are used to update the velocity over a full step; and lastly the displacement is updated again using the latest velocity over another half a step.

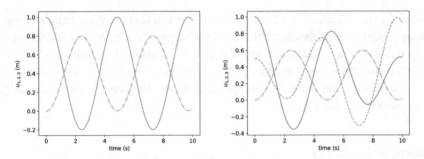

Fig. 6.3 Displacement vs. time of a three-body coupled oscillator for different initial conditions. All initial velocities are zero, but the initial displacements are $u_1, u_2, u_3 = [1, 0, 1]$ m (left) and $u_1, u_2, u_3 = [1, 0, 0.5]$ m (right). In each case, the solid curve corresponds to u_1, the dash-dot curve u_2, and the dashed curve u_3. The u_1 and u_3 curves overlap each other in the left figure due to the initial condition. The parameters used are: $k_1 = k_2 = 1$ N/m, and $m_1, m_2, m_3 = [1, 3, 1]$ kg.

Th similarity between Programs A.3.3 and A.4.2 is not accidental. Because u, v, a are all numpy arrays obeying vector operations, Program A.4.2 effectively simulates an N-dimensional oscillator, $N = 3$ in the present case.

The results from Program A.4.2 corresponding to the special case Eq. (6.13e) are shown in Fig. 6.3 for different initial conditions which determine what normal modes are activated and their mixture. Because all initial velocities are zero, the center of mass is at rest, so there is no translation and only oscillatory normal modes $\mathbf{u}_2$ and $\mathbf{u}_3$ in Eq. (6.14) are expected. The normal frequencies are $\omega_2, \omega_3 = 1, \sqrt{5/3}$ rad/s, respectively.

In the first set of initial conditions (Fig. 6.3 (left)), the two outer masses m_1 and m_3 start with equal displacement, and according to Eq. (6.14), this activates normal mode $\mathbf{u}_3$ only. The displacements u_1 and u_3 remain equal with time (overlapping curves), and oscillate with ω_2. The middle mass also oscillates with the same frequency, but it (u_2) has a different amplitude and is 180° out of phase. From $t = 0$, the outer masses move to the left, the middle mass moves to the right; and just when m_1 and m_3 reach their left-most positions, m_2 is at its right-most position. Then, simultaneously, all three reverse directions and the cycle is repeated all over again. This is the asymmetric stretch. The asymmetric stretch is more clearly seen with visualization (see program in Appendix A).

Note that, though no translation, the center of mass is not at the origin. Therefore, the oscillations are offset from the origin, and the maxima and minima are not symmetric about the horizontal axis.

Now, the results from the second set of initial conditions as shown in Fig. 6.3 (right) are less straightforward to interpret. The middle mass shows a nice, purely harmonic oscillation. However, the oscillations of the outer masses are neither harmonic nor even periodic. To understand this, we note that Eq. (6.14) supports only two normal modes regarding the middle mass, $\mathbf{u}_1$ and $\mathbf{u}_3$, because its displacement is zero in normal mode $\mathbf{u}_2$. As a result, the middle mass can only oscillate with frequency ω_3, regardless of initial conditions.

In contrast, both oscillatory normal modes are open to the outer masses. The initial conditions do not fit any pure normal mode, so both are activated in a mixture of the form $c_2 \sin(\omega_2 t) + c_3 \sin(\omega_3 t)$. Even though each sinusoidal term is periodic, the mixture is not necessarily so, unless the frequencies are commensurate with each other, i.e., their ratio is rational, $\omega_3/\omega_2 = n : l$ with n and l being integers. The actual ratio for the parameters used is $\omega_3/\omega_2 = \sqrt{5/3} : 1$. So the frequencies are not commensurate in this case, and u_1 and u_3 are not periodic as a result.[1] One can also verify this by plotting u_1 vs. u_3 and observing that the path is never closed. Such plots are known as Lissajous figures (see Exercise 6.4).

We have only discussed a special case Eq. (6.13e), and already it shows some interesting, nontrivial features of collective motion. Naturally, we can expect the complexity to increase with more masses. What happens if N is large? In the limit $N \gg 1$, oscillations propagate down a continuum medium, and we need to approach the problem as wave propagation. We consider waves next.

6.2 Wave Propagation

Mechanical waves such as sound waves from the flapping wings of a hummingbird or vibrations on a string propagate in continuum media. This involves collective oscillations of many atoms, and general description of waves should properly fall within the realm of continuum mechanics. Fortunately, the few-body oscillators developed in the previous section are sufficient as a basis for us to model waves in the limit of large N, with minimal change. We do expect, however, the properties of bulk media to enter the model at some point.

6.2.1 The Wave Equation

Just as Newton's second law is necessary to describe the motion of point-like particles, a wave equation governing collective wave motion of many particles is needed here, a form Newton's law for the masses, so to speak.

We approach the continuum wave limit by building an N-body oscillator model approximating a string, shown in Fig. 6.4. Let us imagine dividing a springy string into equal segments of length Δx, and embedding N identical masses m along the string, one for every segment. Let us further assume that m is equal to the mass of one segment so the string may be considered massless. Effectively, we have N masses connected to each other with massless mini springs of length Δx and spring constant k.

---

[1] Even though aperiodic, if one performed a discrete Fourier transform (DFT) on u_1 (or u_3), one would still find only two frequencies in the spectrum. Discussion of DFT or related FFT is beyond our scope (see Ref. Cooley and Tukey (1965)).

x_0 x_{n-1} x_n x_{n+1} x_{N-1}

Fig. 6.4 A string represented by an N-body coupled oscillator model

We now consider the displacement u_n of the mass at x_n along the string.[2] It is part of a three-body oscillator with its neighbors $x_{n\pm 1}$ (Fig. 6.4), so we already know the equation of motion from Eq. (6.15). With only slight change of notation, and rewriting it as a second-order equation, we obtain

$$m\frac{d^2 u_n}{dt^2} = k(u_{n-1} - 2u_n + u_{n+1}). \tag{6.16}$$

We recognize the terms in the parenthesis on the right-hand side are part of a second derivative Eq. (4.4), or more precisely, $u_{n-1} - 2u_n + u_{n+1} = (\Delta x)^2 u''$. Substituting into Eq. (6.16) and dividing both sides by Δx, we find

$$\frac{m}{\Delta x}\frac{d^2 u_n}{dt^2} = k\Delta x\, u_n''. \tag{6.17}$$

Equation (6.17) is almost in the standard wave equation, but first we need to relate the parameters to bulk properties. The ratio $m/\Delta x$ measures the linear mass density, so we set $\rho = m/\Delta x$.

From the discussion following the special case Eq. (6.13b), we saw that cutting a spring in half doubles the spring constant. The same applies here: each time the string is divided, the segments become shorter but the effective spring constant increases such that the product $k\Delta x$ remains constant and finite, namely, $k\Delta x = T$ (T for tension).

A separate analysis enables us to relate T to the bulk properties of the string.[3] Adopting the analysis to each segment of the string ($L = \Delta x$), we have $k = YA/\Delta x$, so $k\Delta x = YA = T$, T being the tension in the string as mentioned earlier. This is the relationship we need.

Because $u_n = u(x, t)$ is a function of independent variables of position and time, we should replace the ordinary derivatives with partial ones, $u'' = \partial^2 u/\partial x^2$ and $d^2 u/dt^2 = \partial^2 u/\partial t^2$. Finally, we rearrange Eq. (6.17) to obtain the wave equation,

---

[2] This is a longitudinal wave where oscillations are in the direction of wave propagation. If the oscillations are in the perpendicular direction, it is called a transverse wave. The discussion assumes the former, but the final result Eq. (6.18) is valid for the latter as well.

[3] Consider a string made up of many thin threads from end to end, each thread acting like a thin spring. The more threads there are, the stronger the string. But as stated above and seen from Eq. (6.13b), the longer the thread, the weaker the thread (hence the string). So the "spring" constant of the string k should be proportional to the number of threads which in turn is proportional to the cross section A of the string, and inversely proportional to its length L, i.e., $k = YA/L$, with Y being a constant. To stretch the string by ΔL, the force from Hooke's law is $F = k\Delta L = YA\Delta L/L$, or $Y = \frac{F}{A}/\frac{\Delta L}{L}$, which is just the Young's modulus. The constant Y is a bulk property of the material.

$$\frac{\partial^2 u(x,t)}{\partial x^2} = \frac{1}{v^2}\frac{\partial^2 u(x,t)}{\partial t^2}, \quad v = \sqrt{T/\rho} \quad \text{(Wave equation)}. \tag{6.18}$$

The wave speed v depends on the tension and mass density. A wave moves faster on a more rigid or lighter string. Equation (6.18) is partial differential equation (PDE) involving two independent variables and requiring special techniques for numerical solutions discussed shortly.

Whilst admiring the simplicity and mathematical beauty of Eq. (6.18), we should also appreciate the physical insight the process revealed. On the one hand, the acceleration of a segment at x is equal to $a = \partial^2 u/\partial t^2$. On the other hand, the force pulling the segment from either the left (F_L) or the right (F_R) neighbors is proportional to the respective differences in displacement (Eqs. (6.1), (4.1)): $F_L \propto [u(x) - u(x - \Delta x)] \propto u'(x)$ and $F_R \propto [u(x + \Delta x) - u(x)] \propto u'(x + \Delta x)$. The *net force* is given by the difference $F_R - F_L = F_{net}$, which turns out to be $F_{net} \propto [u'(x + \Delta x) - u'(x)] \propto u''(x)$ (essentially, the difference of differences is the second derivative; see also Sect. 4.1). Therefore, Eq. (6.18) simply implies $F_{net} \propto a$, Newton's second law in disguise. Incidentally, following the same reasoning, one can show that even though Eq. (6.18) was obtained assuming longitudinal waves, it is equally applicable to transverse waves (see Wang (2016)).

6.2.2 Waves on a String

To simulate waves on a string, we need numerical solutions of the wave equation Eq. (6.18). Unlike the ordinary differential equations of motion (e.g. Eq. (6.15)) we have encountered so far, the wave equation (6.18) is a PDE. It marks a qualitatively different departure from ODEs because a PDE has two independent variables at least whereas an ODE has only one. The standard stepping methods for ODEs no longer work for PDEs, at least without significant reformulation. Another difference is that boundary conditions affect the solutions of a PDE. For instance, waves on a string behave differently when one end is fixed or open. That means numerical methods of solution will need to consider the type of PDEs as well as boundary conditions.

Finite Difference Method

Seeing that the finite time-stepping has served us well in the time domain, it is wholly reasonable to ask whether finite difference in the space domain can be used for solving PDEs like Eq. (6.18). The finite difference method (FDM) is based on this idea: it utilizes discretization of both space and time, and can work quite well for certain problems including waves on a string.

In the FDM, regular grids are set up in both space (Δx) and time (Δt), and differential operations like $\partial^2 u/\partial x^2$ and $\partial^2 u/\partial t^2$ are evaluated over the respective grids. We have already seen this for space with Eq. (4.4), and if we do the same for time, we have

$$\frac{\partial^2 u(x,t)}{\partial x^2} = \frac{u(x - \Delta x, t) - 2u(x,t) + u(x + \Delta x, t)}{(\Delta x)^2},$$ (6.19a)

$$\frac{\partial^2 u(x,t)}{\partial t^2} = \frac{u(x, t - \Delta t) - 2u(x,t) + u(x, t + \Delta t)}{(\Delta t)^2}.$$ (6.19b)

You may notice that the FDM discretization process is essentially the reverse process of deriving Eq. (6.18).

To tidy up notations, we label the space grid with index n as in Fig. 6.4, $x_n = n\Delta x$; and label the time grid with m (not mass) $t_m = m\Delta t$, with $n, m = 0, 1, 2, \ldots$ Let $u_n^m \equiv u(x_n, t_m)$ denote the displacement at point x_n and time t_m. With these notations and Eqs. (6.19a) and (6.19b), the wave equation Eq. (6.18) at $x = x_n$ and $t = t_m$ in FDM turns into

$$\frac{u_{n-1}^m - 2u_n^m + u_{n+1}^m}{(\Delta x)^2} = \frac{1}{v^2} \frac{u_n^{m-1} - 2u_n^m + u_n^{m+1}}{(\Delta t)^2}.$$ (6.20)

We have reduced the wave equation into a difference equation in FDM.

Let us rearrange Eq. (6.20) so the time index m is in decreasing order from left to right,

$$u_n^{m+1} = 2(1 - \beta^2)u_n^m + \beta^2 \left(u_{n-1}^m + u_{n+1}^m\right) - u_n^{m-1}, \quad \beta = \frac{v}{\Delta x / \Delta t}.$$ (6.21)

This is a three-term recursion formula. Its interpretation is straightforward: the displacements (at all space grids) at the next time t_{m+1} can be obtained from those at the current and previous times, t_m and t_{m-1}, respectively. This is just the time-stepping method we are familiar with, but it happens across all space grids.

To get the process started, we need seed values at two steps, u_n^0 at t_0 and u_n^1 at t_1, for all space grids x_n. Because the wave equation is second order in time, we need two initial conditions to uniquely determine the solution.

Finally, we have the important pieces we need to build a FDM code like Program A.4.3 to simulate waves on a string. Executing it, we should see an energetic wave flip-flopping like a butterfly. Snapshots of the wave are shown in Fig. 6.5 at several equal time intervals.

The wave is set up on a string of unit length. One part of the string influences its neighbors and is influenced by them simultaneously, much like the jiggling atomic motion stated earlier. The result is a realistic depiction of wave motion.

Boundary and Initial Conditions

Every point on the string oscillates except the end points. This brings us to the matter of boundary conditions: what to do at the end points of the string? We can see the problem from Eq. (6.21): When $n = 0$ at the left end point, then u_{-1}^m is needed but undefined, and likewise at the right end point.

In fact, there is freedom in the choice of the boundary conditions. The simplest is to fix the ends, i.e., $u_0^m = u_N^m = 0$ (or some other nonzero value, even time-dependent values like a driven oscillator), and we do not need to update the end points. This is our default choice

Fig. 6.5 Waves on a string of unit length at equal time intervals

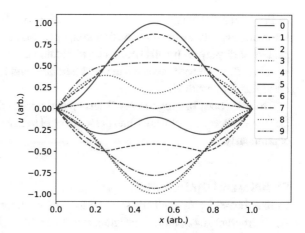

made in Program A.4.3. Other common choices include open ends and periodic boundary conditions (Exercise 6.7).

Having determined boundary conditions, the initial conditions should be chosen consistently. In Program A.4.3, we chose the initial conditions at t_0 as $u^0 = \sin^2(\pi x)$, which is consistent with the fixed-end boundary conditions because they vanish at the end points of a string of unit length, i.e., $u(0, 0) = u(1, 0) = 0$. Any other initial condition satisfying this condition are acceptable. For the immediate next step at t_1, we chose $u^1 = cu^0$. The scaling factor c should be close to one because we assume the time step to be small and the change should be also small for realistic waves. Of course, other choices consistent with boundary conditions are permissible. However, this could result in unrealistic or even unphysical waves.

The main calculation is done in the following module that advances the solution to the next time step.

```
In [8]:  def waves(u0, u1):
             u2 = 2*(1-b)*u1 - u0              # unshifted terms
             for i in range(1, len(u0)-1):
                 u2[i] += b*( u1[i-1] + u1[i+1] )    # left, right
             return u2
```

It is natural to store the wave at a given time in an array of length equal to the number of space grids. On input, the module assumes two arrays u0 and u1 holding the waves at times t_{m-1} and t_m, respectively. Then it computes the wave at t_{m+1}, u2, according to Eq. (6.21): first by including the terms at the space grid x_n (but at different times) with vector array operation, and then looping through interior space grids, adding the contributions from its left and right neighbors. The end grid points x_0 and x_{N-1} are left untouched, and remains zero as required by our choice of boundary conditions.

Program A.4.3 makes use of wave visualization via Matplotlib's animation feature which automatically calls a user-supplied function, `updatefig()` in our case, to update graphical data to be displayed. It calls `waves()` to obtain u2, then swaps the arrays with u0, u1 = u1, u2 to prepare seeds for the next iteration, and finally updates the plot data via the `set_data()` method.

The main body of Program A.4.3 sets up appropriate parameters, the space grids, initial-izes the seeds and the plot area, and passes control to matplotlib's animation module running continuously.

Stability and Optimization

Unlike time-stepping methods used up to now, we have so far not specified the step size Δt in Program A.4.3, and the program runs fine without it. Actually, the step size is woven in the single, dimensionless parameter β in Eq. (6.21). This parameter turns out to be very important to the stability of the FDM.

The reason is that Eq. (6.21) is a three-term recursion formula. Recursions are well known to be numerically either stable or unstable. If stable, a recursion relation is powerful and efficient because errors are reduced with each iteration and do not cause problems. However, if unstable, a recursion relation cannot be used numerically because a small error such as the inevitable roundoff error will grow exponentially, usually ruining the true solution in a few steps.

To check the stability of a recursion relation, one generally performs a von Neumann stability analysis, which is beyond our scope here (see Wang (2016)). We state the result that Eq. (6.21) is stable if $\beta \leq 1$. Furthermore, the solutions are most accurate if $\beta = 1$, the exact value used in Program A.4.3.[4]

Keeping $\beta \leq 1$ is equivalent to implicitly setting a Δt small enough that the FDM remains stable. There is no need to set it explicitly.

As seen earlier, the module `waves()` is responsible for advancing the solutions one step forward. It is called repeatedly. It has an explicit loop running through the interior space grids. Explicit looping is slow in Python as discussed earlier (Sect. 3.3.2), and can significantly downgrade the speed. If speed is important, we can usually replace explicit looping with Numpy's implicit looping (Sect. 3.3.2), or optimize in other ways. The drawback is that the former, though efficient, can make the code less understandable sometimes, and the latter does not always lead to a significant speedup.

Still, it is always a good idea to have a working program that may be slow rather than a fast program that may not work. In Program A.4.3 we chose the latter with Numba opti-mization (Sect. 3.6). We simply decorate the module with the `@jit` command. Speed tests (Exercise 6.6) showed that both Numpy's implicit looping and Numba optimization were

---

[4] There is an interesting connection to a broader array of phenomena including the uncertainty prin-ciple (see Sect. 7.2).

about 12 times faster than explicit looping. Considering the ease of use, Numba optimization is the clear choice here.

6.2.3 Fundamental Modes

Just like there are characteristic frequencies (normal modes) in few-body oscillations (Sect. 6.1.1), we expect similar fundamental frequencies (harmonics) to exist in waves on a string.

To find them, let us resort to perhaps the most useful method for solving PDEs: the separation of variables. We propose a trial solution consisting of a product of space and time variables separately,

$$u(x, t) = X(x)Y(t). \tag{6.22}$$

The method is successful if we can reduce the wave equation Eq. (6.18) into separate equations, each involving only one variable x or t.

To that end, we substitute the trial solution into Eq. (6.18) and rearrange it such that x and t appear alone on each side,

$$\frac{1}{X}\frac{d^2 X}{dx^2} = \frac{1}{v^2}\frac{1}{Y}\frac{d^2 Y}{dt^2}. \tag{6.23}$$

Because x and t can vary independently, Eq. (6.23) can hold if and only if each side is a constant for all x and t, namely,

$$\frac{1}{X}\frac{d^2 X}{dx^2} = -k^2 = \frac{1}{v^2}\frac{1}{Y}\frac{d^2 Y}{dt^2}, \quad \text{or}$$
$$\frac{d^2 X}{dx^2} = -k^2 X, \quad \frac{d^2 Y}{dt^2} = -k^2 v^2 Y, \tag{6.24}$$

where we set the separation constant to be $-k^2$ for convenience. Now we just need to deal with two ODEs.

Though we know the analytical solutions to Eq. (6.24) are sinusoidal from earlier discussions Eq. (5.16), we wish to study it numerically here. Expressing the second derivative on the space grid as in FDM (Eq. (6.20) earlier, and letting $X_n \equiv X(x_n)$, we have

$$\frac{X_{n-1} - 2X_n + X_{n+1}}{(\Delta x)^2} = -k^2 X_n, \quad n = 0, 1, 2, \ldots, N - 1. \tag{6.25}$$

These are N linear equations.

Now using the fixed-end boundary conditions $X_0 = X_{N-1} = 0$ we express the remaining $N - 2$ interior equations implied by Eq. (6.25) in matrix form as (Exercise 6.5),

$$
\mathbf{AX} = k^2\mathbf{X}, \quad \mathbf{A} = \frac{1}{(\Delta x)^2}
\begin{bmatrix}
2 & -1 & & & \\
-1 & 2 & -1 & & \\
& \ddots & \ddots & \ddots & \\
& & -1 & 2 & -1 \\
& & & -1 & 2
\end{bmatrix}, \quad
\mathbf{X} =
\begin{bmatrix}
X_1 \\
X_2 \\
\vdots \\
X_{N-1} \\
X_{N-2}
\end{bmatrix}
\tag{6.26}
$$

This is just like the eigenequation seen earlier Eq. (6.5). The system matrix $\mathbf{A}$ is square $(N-2, N-2)$ and tridiagonal, with nonzero elements only on the main diagonal and one row or column immediately off of it.

We expect Eq. (6.26) to yield eigenvalues and eigenvectors just as Eq. (6.5) does. Rather than solving it with symbolic computation because it is impractical for even a moderate N, we aim to solve it numerically, which turns out to be just as easy with the Scipy library (*see* Section 3.5). The implementation is given in Program A.4.4.

The main task is the preparation of the system matrix $\mathbf{A}$:

```
In [9]:  A = np.diag([2.]*(N-2))                              # diagonal
         A += np.diag([-1.]*(N-3),1) + np.diag([-1.]*(N-3),-1)  # off diags
         A = A/(x[1]-x[0])**2           # delta x = x_1 - x_0
         lamb, X = eigh(A)              # solve for eigenvals and eigenvecs
```

This task is made easy with Numpy's `diag(values, offset)` function (Sect. 3.3.2), which fills a diagonal with `values` (array). The `offset` parameter (default 0) is the distance from the main diagonal, so ±1 is just what is needed for a tridiagonal matrix $\mathbf{A}$. Once $\mathbf{A}$ is constructed, the eigenvalues and eigenvectors (fundamental modes) may be obtained from the Scipy `eigh()` module.

Figure 6.6 shows the first four fundamental modes from Program A.4.4. They are standing waves because the nodes remain at zero at all times. Each mode looks like parts of a sinusoidal wave, all being zero at the boundaries.

Fig. 6.6 Standing waves on a string of unit length. Only the lowest four fundamental modes are shown

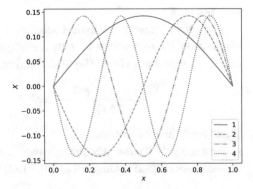

Allowing for a phase, the sinusoidal waves can be written as

$$X = a \sin(kx + \phi). \tag{6.27}$$

Figure 6.6 suggests $X(0) = X(1) = 0$, namely

$$\sin(\phi) = \sin(k + \phi) = 0. \tag{6.28}$$

The first term in Eq. (6.28) requires $\phi = 0$, reducing the second term to $\sin(k) = 0$. So we must have $k = m\pi$, and $m = 1, 2, \ldots$, that is, only certain, discrete k_m values are allowed. Conversely, the wavelength is the period in x when $k_m \lambda_m = 2\pi$, so

$$\lambda_m = \frac{2}{m}, \quad m = 1, 2, 3 \ldots \quad \text{(wavelength on unit-length string)} \tag{6.29}$$

These are the fundamental wavelengths. Both Eq. (6.29) and Fig. 6.6 show that only multiples of half-waves can fit (see also Sect. 7.3.1). Fundamental frequencies are also discrete and related to the wavelength by $f_m = v/\lambda_m = mv/2$. Incidentally, the quantity $k_m = 2\pi/\lambda_m$ is also known as the wave vector and is inversely proportional to the wavelength.

The computed k values are equal to the square root of the eigenvalues from Program A.4.4. The lowest four are $k/\pi = [0.99995888, 1.99967103, 2.99888979, 3.99736862]$. These numbers are accurate to their respective exact integer values within an error less than 0.01% with $N = 101$ grid points. The accuracy will increase with larger N.

Finally we note two things about fundamental modes. First, fundamental modes are the basic building blocks of waves. Any wave such as the one shown earlier (Fig. 6.5) can be written as a superposition of these modes. Second, fundamental modes are discrete because of boundary conditions. We will see discrete eigenvalues in the allowed energies later in quantum mechanical problems precisely due to the same requirements.

6.3 Exercises

6.1. Initial conditions. Given the initial positions and velocities of the two particles, derive the constants shown in Eq. (6.11). Consider setting up the equations and solve them using Sympy (Sect. 3.3).

6.2 Two-body oscillator. Two-body oscillations may be decomposed in terms of fundamental modes Eqs. (6.8) and (6.9). The following activities illustrate the process.

(a) Determine the constants Eq. (6.11) from the initial conditions given in Fig. 6.2. Now we have the fundamental modes Eqs. (6.8) and (6.9), and the displacements $u_1(t)$ and $u_2(t)$ (first and second rows of Eq. (6.10), respectively). For instance, $u_1(t) = u_{cm} + v_{cm}t - \frac{m_2}{m_1} A \sin(\omega_2 t + \phi)$.

(b) Next, plot the displacement from a simulation with the same initial conditions using Program A.4.2, where we set $k_2 = 0$ because we need a two-body oscillator.

Also plot the fundamental modes from part 6.2a. You should have a graph similar to Fig. 6.2. How close are the two results?

*(c) Repeat part 6.2b but for the velocities. Discuss similarities and differences in comparison to results for displacement.

6.3. Coupled three-body oscillator.

(a) Write down the equations of motion for a three-body coupled oscillator (Fig. 6.1 (b)) similar to Eq. (6.2a) and (6.2b). Set up the system matrix **A**, and find the eigenvalues symbolically with Sympy, Eq. (6.12). You need to simplify the result further manually using reduced masses μ_{ij} that Sympy does not recognize without our help.

(b) From the general expression above, verify the special cases in Eqns. (6.13a) to (6.13e). For the special case Eq. (6.13e), find the eigenvectors Eq. (6.14), and sketch the motion represented by the normal modes $\mathbf{u}_2$ and $\mathbf{u}_3$.

6.4. Lissajous figure.
Graph the displacements u_1 vs. u_3 corresponding to the initial condition of Fig. 6.3 (right). Use Program A.4.2 as a template, and increase the end time of the loop so that many sweeps are plotted. Is the path closed? Comment on why. Next, choose a different initial condition and repeat. Lastly, find a mass ratio for m_1/m_3 where you expect to see commensurate frequencies, and generate new Lissajous figures. Are the paths closed?

6.5. Fundamental modes.

(a) Verify the system matrix **A** in Eq. (6.26). Write down explicitly the second, third, and the next to last rows of Eq. (6.25) ($n = 1, 2, N - 2$, respectively) to see the pattern.

(b) Overlay the fundamental modes X_m (Eq. (6.27) with λ_m from Eq. (6.29)) on the same plot as the numerical ones from Program A.4.4. Normalize them at the peak. Comment on the match.

(c) One important property of fundamental modes is that they are "orthogonal" to each other, namely, the overlap integral $\int_0^1 X_i X_j \, dx = \delta_{ij}$, with $\delta_{i=j} = 1$ and $\delta_{i \neq j} = 0$, also known as the Kronecker delta function.

Calculate the overlap integrals for the lowest modes shown in Fig. 6.6. Use Simpson's rule (Sect. 4.2) because the eigenvectors from Program A.4.4 are on a regular grid. Comment on the results. Note that integrals are not necessarily unity for $i = j$ because the eigenvectors are not normalized.

6.6. Speed optimization. Speed can be an issue in larger calculations. In wave propagation simulations (Program A.4.3), for instance, it can be slow with larger grid size or longer propagation time. Let us check code execution times and with optimization.

(a) Time the execution of the main module `waves` in Program A.4.3 with and without Numba optimization by leaving the `@jit` decoration in or commenting it out. Execute the wave updating function `waves` some large number of times (say 10^5) to obtain accurate timing (see Sect. 3.6 and Program A.1.8). Compute the speedup factor.

*(b) Replace explicit looping in `waves` with implicit looping with Numpy arrays (Sect. 3.3.2). Note the left and right grid points (middle terms in Eq. (6.21)) can be obtained with index slicing. Time the code again and compute the speedup factor. Compare with results above.

6.7. Wave propagation. We explore several aspects of wave propagation more deeply here including standing and traveling waves, boundary conditions, and energy in waves. The template code for this exercise is Program A.4.3.

(a) Standing waves have fixed nodes where the displacement remains zero at all times. To generate a standing wave, change the initial conditions to $u(0) = \sin(2\pi x)$ and $u(\Delta t) = cu(0)$ (c should be close to 1, say 0.98). Run the simulation and describe the results.

If we want a standing wave with two nodes (excluding the end points), how do you set the initial conditions? Write down your hypothesis. Run the code and test the hypothesis.

(b) Next, generate a traveling sine wave. A traveling wave refers to the waveform (shape of the wave) moving along the direction of travel and keeping the same (or nearly the same) shape (translational motion). Based on your understanding from making standing waves earlier, describe how you would change $u(\Delta t)$ to make a traveling wave. Again write down your plan. Implement the plan and test it. Produce a few snapshots of the wave.

Did your plan work? If so, explain your reasoning. If not, revise your plan and try again.

*(c) Besides fixed end points, there are other common choices such as open ends and periodic boundary conditions (PBC). The latter is easier to implement so let us examine that.

In PBC, we assume the string is infinite long and we are looking at one segment of it only. This means that there are identical copies of the grids to the left and to the right of our segment. If we are at the left end point u_0, the neighbor immediately to the left, u_{-1}, would be equivalent to u_N, the right end point. Conversely, if we are at the right end point, the point immediately to the right, u_{N+1}, would be equal to the left end point, u_0. Together, we have the following for PBC,

$$u_{-1} = u_N, \quad u_{N+1} = u_0. \tag{6.30}$$

Modify Program A.4.3 to use PBC. Describe the difference in the behavior of the waves.

*(d) Waves carry energy. There are both kinetic energy due to the jiggling of each point and potential energy due to the elasticity of the string (spring).

The kinetic energy may be calculated directly as follows. Consider a small segment at grid n of width Δx, and displacement velocity v_n. The total kinetic energy is the sum from all segments, $K = \sum_n \frac{1}{2} \rho \Delta x v_n^2$, with ρ being the mass density of the string. The velocity v_n at time t_m can be calculated using the change in displacement with the midpoint approximation Eq. (4.3) as $v_n = (u_n^{m+1} - u_n^{m-1})/2\Delta t$ (see Eq. (6.20)). One may also use the two-point formula Eq. (4.1), at least initially when we have only two points in time.

Substituting v_n into K, show that the final result is

$$K = C \sum_n \left(u_n^{m+1} - u_n^{m-1} \right)^2, \quad C = \frac{1}{2} \frac{\rho^3}{\beta^2 \Delta x}, \tag{6.31}$$

where β is given in Eq. (6.21). Verify that K has the dimensions of energy. Note that Δt is absorbed in β.

*(e) Now compute and graph the kinetic energy Eq. (6.31). We need be concerned with the constant C if we are interested in only how the kinetic will change, or equivalently, the ratio $\kappa = K/C$. Calculate κ by modifying Program A.4.3. First, write a function to calculate the kinetic energy given the displacements at t_{m-1} and t_{m+1}. Second, introduce a loop to propagate the wave and record the kinetic energy for some time. Finally, graph the result after the loop as a function of time (iteration number m). You can remove or comment out the animation part of the main function to speed things up. Here is the pseudo code.

```
# kinetic energy at t
def KE(u0, u2):     # u0, u2 = displacements at t−dt and t+dt
    for every grid point n:
        cumulate sum (u2[n+1]−u0[n−1])**2
    return sum

while t < tend:   # or use time index m
    update u2 from u0 and u1
    call KE
    append t and KE
    increment t
    reseed u0, u1

comment out animation
plot KE list vs t list
```

Discuss the result and features. Note the small oscillations in the curve. Are they real or numerical artifacts? How can you check?

On average, the potential energy is equal to the kinetic energy, so the latter gives us an idea of the total energy in the wave.

References

J. W. Cooley and J. W. Tukey. An algorithm for the machine calculation of complex Fourier series. *Math. Comp.*, **19**: 297–301, 1965.

G. R. Fowles and G. L. Cassiday. *Analytical mechanics*. (Thomson Brooks/Cole, Belmont, CA), 7th edition, 2004.

J. Wang. *Computational modeling and visualization of physical systems with Python*. (Wiley, New York), 2016.

Modern Physics and Quantum Mechanics 7

Physics stood at the cusp of dramatic change near the end of the nineteenth century and the beginning of the twentieth century. Prior to that, there was a sense that physics was mature, complete and well-established, and broad subjects like mechanics, electromagnetism, and thermodynamics, which we call classical physics now, captured all the big ideas necessary to understand nature. It was as if all that was left to do was to work out some details of physics.

Then, strange things were discovered that were incompatible and could not be readily explained with classical physics, and the sense of physics being complete was shattered. A patchwork of ideas like a jigsaw puzzle was proposed to address various shortcomings of classical physics, including attempts to explain the constancy of speed of light, the ultraviolet catastrophe, the discrete atomic spectra, and so on. But none of the patchwork gave a coherent, complete, or satisfactory explanation of new discoveries. More important, it was unable to make new, testable predictions. Out of these chaos came two main theories, relativity and quantum mechanics, which we now call modern physics. Modern physics completely changed our understanding of space-time and laws of physics at the microscopic level. We will discuss simulations of select topics of modern physics next.

7.1 Relativity of Space and Time

Oftentimes there are certain implicit assumptions in physics that we take for granted to be self-evident and true. The concept of space and time in classical physics is a prime example. Newtonian physics assumes space and time to be absolute across different reference

© The Author(s), under exclusive license to Springer Nature Switzerland AG 2023   139
J. Wang and A. Wang, *Introduction to Computation in Physical Sciences*, Synthesis Lectures on Computation and Analytics,
https://doi.org/10.1007/978-3-031-17646-3_7

frames and observers.[1] This assumption seems natural and eminently reasonable, and more important, it works well within the confines of Newtonian physics. However, mounting experimental evidence at the of twentieth century led to the realization that space and time are relative.

7.1.1   Einstein's Postulates

Consideration of the speed of light is at the heart of the theory of relativity. Being a wave, light propagation is described by Maxwell's electromagnetic wave equation analogous to waves on a string Eq. (6.18) with displacement u replaced by the electric field vector $\mathbf{E}$, $\partial^2\mathbf{E}/\partial x^2 = c^{-2}\partial^2\mathbf{E}/\partial t^2$, where c is speed of light (similar to v in Eq. (6.18)). Unlike the mechanical wave speed that depends on the tension and density, c is equal to $c = 1/\sqrt{\mu_0\epsilon_0}$, where μ_0 and ϵ_0 are the respective permeability and permittivity constants of free space. This means that c is the same for every observer in free space.

Let us consider the implication in a thought experiment with two observers, Stacy on a stationary platform and Moe on a train moving with velocity v in the positive x direction. Suppose Stacy throws a fast ball with velocity u in the direction of train motion at the moment Moe passes her. The ball hits a pillar at a distance d from Stacy after a time interval Δt, so $d = u\Delta t$. From Moe's perspective on the train, he sees the ball travel a shorter distance because the train has moved a distance $v\Delta t$, so $d' = d - v\Delta t$. Moe then calculates the velocity of the ball relative to him as $u' = d'/\Delta t = u - v$, exactly as is expected in Newtonian mechanics.[2]

Now, instead of a ball, what if Stacy shines a laser light? The situation would seem to be analogous: Stacy would see speed of light to be c, and Moe $c' = c - v$ (also see Exercise 7.2). But there is a problem: the speed of light would be different to different observers, contradicting the conclusion that it be the same for everyone. To explain this apparent contradiction, the *ether* hypothesis gained considerable traction. It presumed the existence of a particular medium called ether. In a reference frame where ether was at rest, the speed of light would be c. In any other reference frame, there would be a ether wind slowing down the speed of light. Elaborately designed experiments had been performed, exemplified by the famous Michelson-Morley setup, to look for ether. All yielded a null result, ether was never found.

One reason we have the apparent contradiction is the implicit assumption of the time interval Δt being the same in both reference frames. Einstein's postulates of relativity fundamentally change the concept of space and time of classical physics. Simply stated,

---

[1] Some would propose, light heartedly of course, but only slightly, that there ought to be an additional, Newton's zeroth law stating explicitly that space and time are absolute, and the positions and velocities of particles are well-defined simultaneously so the concept of a trajectory is valid.

[2] Such a relation is also known as a Galilean transformation.

- All observers in initial reference frames (non-accelerating) are equivalent to each other.
- Speed of light is constant in all inertial reference frames.

The first postulate states that no observer can claim to be special, and the second one just states an experimental fact. Each of the statements is eminently reasonable by itself, but together, they lead to profound results. We simulate one such result next.

7.1.2   Time Dilation and Length Contraction

Armed with Einstein's postulates, we discuss now, and simulate afterward, a slightly different thought experiment than the one described earlier. Now, Moe fastens on the floor of the train a laser pointing up toward the ceiling at a height h above. He also attaches a mirror to the ceiling facing downward and directly above the laser.

At the moment Moe passes Stacy, he switches on the laser going vertically up toward the ceiling. Upon reaching the mirror, the laser is reflected back down and eventually returns to the floor. To Moe, the round-trip time of travel of the laser is $\Delta t' = 2h/c$.

Let us consider the same events from Stacy's perspective on the platform. Because the train is moving, Stacy will first see the laser's path moving diagonally up during the ascending phase, and then diagonally down during the descending phase. The process is animated in Program A.5.1 and results are shown in Fig. 7.1.

In the elapsed time Δt on the platform, the train will have moved a distance $v\Delta t$. The total path length of the laser is $d = 2\sqrt{h^2 + (v\Delta t/2)^2}$. Because speed of light is also c

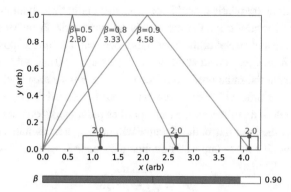

Fig. 7.1 Paths of a light beam as seen on the stationary platform in a thought experiment (see text). The time intervals as measured on the platform and on the train at different velocities are given at the top and bottom, respectively, in units of h/c where h is the height of the train car. The length of the box at the bottom depicts qualitatively the length of the train car as seen by an observer on the platform

in Stacy's reference frame (platform), she will conclude her time interval $\Delta t = d/c = 2\sqrt{h^2/c^2 + (v\Delta t/2c)^2}$.

Stacy and Moe will have measured different time intervals between the events. They are related as

$$\Delta t = \frac{\Delta t'}{\sqrt{1 - \beta^2}}, \quad \beta = \frac{v}{c}. \tag{7.1}$$

When they compare the recorded times, they will readily agree that the times are indeed different and that each made the measurement applying the same laws of physics, i.e., the same definition of time. They conclude that the difference is due to time being relative rather than absolute between reference frames.

Equation (7.1) requires $v \le c$ for both times to be real, so c is the speed limit of the cosmos. Because the denominator in Eq. (7.1) is less than one, it follows that $\Delta t > \Delta t'$, namely, the moving clock (Moe's) runs slower than a stationary clock (Stacy's). This is called *time dilation*.

We can model the light path and time dilation effect with Program A.5.1. The time is measured in units of h/c, so the round-trip time is always $\Delta t' = 2$ in Moe's reference frame. With increasing β, both the path length and time dilation increase in Stacy's reference frame. However, varying β with the slider, time dilation becomes noticeable only for $\beta \gtrsim 0.1$, or $v \gtrsim 3 \times 10^7$ m/s ($\beta \sim 0.15$ would yield 1% difference). This is a very large speed rarely encountered in classical physics, about 100,000 times faster than speed of sound in the air. Relativistic effects are insignificant in everyday-life activities where Newtonian mechanics works very well and can be regarded as the zeroth-order approximation of relativistic mechanics. However, they are significant at high speeds such as in motion of elementary particles where speed can approach c.

If motion and time are relative, could not Moe also claim time dilation of Stacy's measurement? Lest there is confusion, the situation is not symmetric. From Moe's point of view, the events, departure and arrival of the laser pulse, occur at the same x' position (origin). In Stacy's reference frame, they occur at different positions, $x = 0$ and $v\Delta t$. The time interval of events occurring at the same position is known as the *proper time*. Only Moe can claim to measure the proper time. Other observers will measure dilated time (see Exercise 7.1).

Let us look at relativity of space. If Stacy splashes paint marks on the platform between the start and end of the laser pulse, the distance will be $L = v\Delta t$ as stated earlier. Moe will see the platform moving in the opposite direction a distance $L' = v\Delta t'$. Using Eq. (7.1) the two distances are related as

$$L' = \sqrt{1 - \beta^2} L. \tag{7.2}$$

Moe finds a shorter length, an effect called *length contraction*. Note that Stacy's measurement of L is with the platform at rest. Length of an object at rest is called the *proper length*. Other observers like Moe will measure a contracted length, i.e., moving objects are shorter (see Exercise 7.2).

Now turn it around from Stacy's point of view. The length of the train car (proper length) will appear contracted to her. Figure 7.1 shows increasing amount of contraction with increasing train velocity.

7.2   Diffraction and Double-Slit Interference

Interference is a hallmark of wave phenomena and can be observed in everyday life. For instance, the color bands we see on a soap bubble or on an oily water surface are due to interference between light reflected from the top and bottom of the thin layer, and the uneven layer thickness causes constructive interference to occur at different spots for different colors.

For interference to occur, two or more coherent sources are required. The classic Young's double-slit setup has two slits separated by a distance d, splitting an incoming wave into two such coherent emitters which then interfere constructively or destructively, creating bright and dark patterns on a screen behind the slits. The double-slit interference is especially important in understanding the wave nature of particles in modern physics.

When the incoming wave front reaches the slits, each single slit acts like a system of secondary sources emitting outgoing spherical waves (Fig. 7.2). Conceptually it is helpful to imagine the slit divided into N secondary sources along the width of the slit. Then we can model the spherical wave from the jth source as

$$\psi_j = \frac{e^{ikr_j}}{r_j}, \quad r_j = (r^2 + x_j^2 - 2rx_j \sin\theta)^{\frac{1}{2}}, \quad k = \frac{2\pi}{\lambda}. \tag{7.3}$$

In Eq. (7.3), x_j is the (horizontal) position of the jth secondary source relative to the origin (or center in the case of a single slit), and x the position on the screen (Fig. 7.2). The wave vector k is inversely proportional to the wavelength λ as usual (Sect. 6.2.3).

The advantage of spherical waves is that they preserve flux. Though plane waves (without $1/r_j$) are simpler to use analytically, spherical waves are just as easy in numerical compu-

Fig. 7.2 The geometry of a single-slit emitter. The vertical arrows represent the incoming wave front. The position vectors indicate the paths of secondary waves emitted from different parts of the slit converging on the screen. The width of the slit is w, and the distance to the screen is h

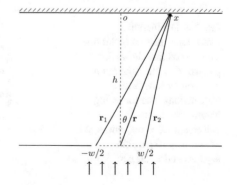

tation. We note that, because the single-slit width is usually small compared to the screen distance $w \ll h$, both types of waves produce similar results.

To find the interference pattern, we add up all the secondary waves from each slit, e.g., $\Psi_1 = \sum_j \psi_j$ from slit 1, and similarly Ψ_2 from slit 2 (taking into account separation d). The total double-slit interference pattern I (intensity) is given by

$$I = |\Psi|^2 = |\Psi_1 + \Psi_2|^2. \tag{7.4}$$

A program to calculate interference patterns from Eq. (7.4) just amounts to summing up all the spherical waves Eq. (7.3) and is straightforward to write. Our implementation is given in Program A.5.2. Apart from the relatively simple task of performing the sum, the bulk of the program deals with using Matplotlib widgets such as sliders for changing the parameters and animating the interference pattern. Representative results are shown in Fig. 7.3.

We see symmetric patterns about the center of the screen, showing peaks from total constructive interference and valleys from total destructive interference. The phase differences due to the path differences of the waves arriving at a given spot on the screen are responsible for the patterns. When the phase differences are zero or multiples of 2π, the waves are in phase and they interfere constructively (bright spots). Conversely, when the phase differences are odd multiples of π, the waves are out of phase and they interfere destructively (dark spots). The first dark spot on either side of the center peak occurs at $\sin \theta_1 = \lambda / w$. With the parameters used in Fig. 7.3, this corresponds to positions $x \sim \pm 0.06$m on the screen (Exercise 7.3).

Figure 7.3 also shows that the interference peaks are modulated by the single-slit diffraction envelope. We can see the latter by dragging the slit-separation slider to zero. Now we have effectively a single slit, giving rise to the broad diffraction envelope matching the double-slit interference profile. When the separation is finite, the slits act as two secondary sources. When the waves from these sources reach the screen and subsequently interfere, the envelope of single-slit intensity distribution remain. Also, with increasing slit separation, path differences increase with angle θ (Fig. 7.2), and the number of peaks increases and also occur more densely (see Exercise 7.4).

Fig. 7.3 Double-slit interference pattern (solid line). The dashed line shows the pattern when the separation between the slits is zero (diffraction). The following parameters are used: wavelength $\lambda = 600$ nm, width of slit $w = 10$ μm, slit separation $d = 50$ μm

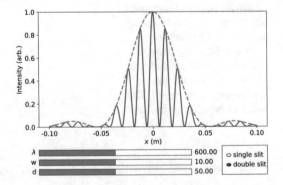

Double-slit interference was a well-understood wave phenomenon, so it was quite a puzzle when elementary particles such as electrons were observed to produce similar interference patterns. The inescapable conclusion was that the particles behaved like waves. But we classically associate properties like mass, momentum and energy with particles, not wavelike things like wavelength.

On the other hand, photoelectric effect demonstrates that light waves also behave like particles, or photons, with energy $E = hf$ where h is the Planck's constant and f the frequency. It is instructive then to relate particle- and wave-like properties via photons. Einstein's energy-momentum relation states $E = \sqrt{m^2c^4 + p^2c^2}$. Photons have no rest mass ($m = 0$) because there is no such thing as a photon at rest. That leaves $E = pc = hf$, and with $f = c/\lambda$, we have for photons,

$$\lambda = \frac{h}{p}. \quad \text{(de Broglie wavelength)} \tag{7.5}$$

It turns out Eq. (7.5) is equally valid for particles of nonzero mass. It is known as the de Broglie wavelength, and can be associated to the wavelike property of any particle. Sometimes we use the term matter wave to describe particles exhibiting wave behavior.

According to Eq. (7.5), by choosing the wavelength of particles, say electrons, with appropriate energy (hence momentum), diffraction and double-slit interference effects exactly analogous to light can be created. Because only whole, not fractional, electrons are observable, the interference pattern consists of density distributions of electrons. Consideration of diffraction and interference by particles leads to several important consequences discussed next.

Uncertainty Principle

Let us assume a diffraction pattern as depicted in Fig. 7.3 (outline) is made with particles such as electrons. Most of the distribution is contained within the central peak, so the diffracted particles have an angular uncertainty roughly between the first minima on either side of the peak.

The angular uncertainty is a direct result of the particles passing through the slit. Before incident on the opening, the particles have a well-defined momentum (vertically, see Fig. 7.2) but not well-defined position. When we restrict the particle to within the slit, we know the position within an uncertainty $\Delta x = w$. However, in doing so, we do not have a well-defined momentum anymore, because the diffracted particle now has an uncertainty in horizontal momentum $\Delta p_x = p \sin \theta$. The product of the uncertainties in position and momentum is (see Exercise 7.5)

$$\Delta x \Delta p_x \simeq 4\pi \hbar \geq \frac{\hbar}{2}, \quad \hbar = h/2\pi. \tag{7.6}$$

The Heisenberg uncertain principle is a fundamental theorem in quantum mechanics stating that $\Delta x \Delta p_x \geq \hbar/2$. It may seem shrouded in mystery to beginning learners of quantum mechanics. Diffraction clearly satisfies the uncertain principle, but also shows us how it comes about. To the extent that particles behave like waves, any attempt to measure its position will result in an uncertainty in momentum, and vice versa. Measurement requires some way of probing the particle, and this act will cause a disturbance and hence changes its state. Diffraction is an illustrative example. Let us look at measurement in double-slit interference.

The Measurement Problem

Again, let us assume the double-slit interference is created with electrons. If the interference pattern is made up of individual-particle buildup, how does each electron choose where to end up, one may wonder? Are the electrons correlated in some predetermined way? To help answer this question, we might try to send one electron at a time through the slits. A strange thing happens: the interference pattern may take longer to build up, but it is the same nonetheless. This suggests that an electron interferes with itself, i.e., it is as if an electron could be at two places (slits) simultaneously.

Suppose we insisted on knowing which slit the electron came through. Without blocking any slit, we could place a detector closely behind one of the slits so that if an electron is detected, we will know where it came from. But an even stranger thing happens: such a detection scheme would show no interfere pattern at all. In other words, measuring where an electron came from seems to change the state of the electron so much as to totally destroy interference. The act of measurement induces what is known as the collapse of the wave function. Prior to the measurement, the electron wave is a coherent sum (superposition, Eq. (7.4)) of both slits. After the measurement, it ceases to be a superposition. Such measurements amounts to transforming the coherent sum Eq. (7.4) of wave amplitudes into an incoherent sum of intensities $|\Psi_1|^2 + |\Psi_2|^2$ without phase information, so naturally one would not expect interference to occur. In practical terms, it is unnecessary to know where the electron came from for making predictions, and so the question seems to be ill-posed. However, this so-called quantum measurement problem is still not satisfactorily explained to date. It is the main reason why quantum mechanics remains rigorous mathematically but mysterious conceptually. More discussion on measurement follows later (Sect. 7.4).

7.3   Visualizing Quantized States

The preceding discussion barely scratches the surface of the conceptual intricacies in quantum mechanics, a fascinating subject that still has an aura of mystique even though it is over a century old. Among many puzzling aspects faced by a typical beginner in quantum mechanics, the fact that only discrete energy states, as opposed to continuous energies, are allowed is one of the most unsettling concepts. It seems as if this was a foreign, and somewhat arbitrary, concept that one must accept to get anywhere in quantum mechanics. Part of the reason certainly has to do with mathematical details obscuring the bigger picture, and unfamiliarity of some beginners with boundary conditions and how they are applied.

7.3.1 A Quantum Particle in a Box

Here we take a visual approach to understanding the concept of discrete, or quantized, states. The idea is to see how quantized energy comes graphically and naturally from reasonable requirements of the solutions, the same ones to obtain waves in classical mechanics such as waves on a string.

We choose the simplest model system known as the particle in a box. Imagine a particle confined between two rigid walls separated by a distance a. The particle can move freely along a straight line between the walls (no potential, $V(0 < x < a) = 0$), but it cannot escape outside of the hard walls ($V = \infty$ for $x < 0$ or $x > a$). The classical picture is a particle bouncing between the walls.

Quantum mechanically, the state of the particle is governed by the Schrödinger equation (see Griffiths and Schroeter (2018)),

$$\frac{d^2\psi(x)}{dx^2} = \frac{2m(V - E)}{\hbar^2}\psi(x) = -k^2\psi(x), \quad k = \sqrt{2mE}/\hbar, \quad (0 \leq x \leq a, \ V(x) = 0)$$
(7.7)

where $\psi(x)$ is the wave function describing the state of the particle, E the energy of the particle (positive), m the mass of the particle, and $\hbar = h/2\pi$ with h being Planck's constant as before Eq. (7.5). In Eq. (7.7), we assume one wall is at the origin and the other at $x = a$.

The wave function $\psi(x)$ is very important because every property and observable of the system can be determined from it. Paradoxically, $\psi(x)$ itself cannot be directly measured, only $|\psi(x)|^2$, the probability density of finding the particle. This is similar to the intensity of waves Eq. (7.4) discussed earlier in interference. A reasonable and physically acceptable wave function must be continuous throughout. Otherwise, we would find different probabilities of detecting a particle depending whether the detector is placed to the left or the right of a point in space, an unphysical outcome.

We can verify that $\sin(kx)$ and $\cos(kx)$ are solutions of Eq. (7.7) with the potential $V = 0$, so is a linear combination of the two,

$$\psi(x) = A\sin(kx) + B\cos(kx), \quad (0 \leq x \leq a)$$
(7.8)

where A and B are constants. We take ψ to be a general solution, which should be continuous throughout space.

Because the particle is confined to within the walls, the wave function must be identically zero outside, namely, $\psi(x) = 0$ for $x < 0$ or $x > a$. The continuity requirement just discussed means that the inside wave function should connect continuously to the outside wave function at the walls,

$$\psi(0) = \psi(a) = 0. \quad \text{(continuity)} \tag{7.9}$$

The continuity condition is also known as the boundary condition.

7.3.2   Fitting a Wave Function to the Box

We know the solution Eq. (7.8) satisfies the Schrödinger equation Eq. (7.7), but not necessarily the continuity condition Eq. (7.9). To see this, we plot $\psi(x)$ in Fig. 7.4a with a set of arbitrary constants A, B, and the parameter k, where we rewrite k in terms of $k = n\pi/a$ for reasons that will become clear shortly. The wave function is continuous and smooth, but is not zero at $x = 0$ and a, so it is not an acceptable wave function satisfying Eq. (7.9).

We can choose to determine the constants and parameters mathematically such that the continuity condition is satisfied (see below). However, we choose to do this directly with Program A.5.3 which lets us adjust the parameters interactively with sliders and visually

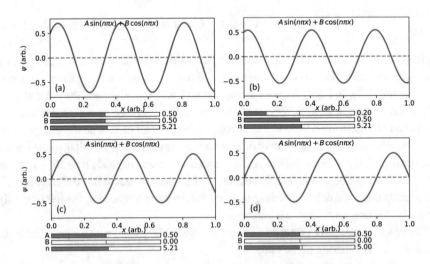

Fig. 7.4 The wave function ψ in an infinite potential well as a function of the scaled variable x/a. The output is from Program A.5.3 with different values of A, B, and $n = k/\pi$. (a) A general solution not satisfying the continuity condition, $\psi(0) \neq 0$ and $\psi(a) \neq 0$. (b) Adjustment of A primarily affects the amplitude of ψ, not the end values. (c) Adjustment of B is the only way to change $\psi(0)$, and only when $B = 0$ the continuity at $x = 0$ is satisfied, $\psi(0) = 0$. (d) Adjustment of n changes the number of oscillations. Whenever n is an integer, however, continuity is satisfied at the right end $\psi(a) = 0$

inspect how the wave function changes. The results are shown in Fig. 7.4 for representative sets of parameters.

For arbitrary values of the parameters, the wave function does not satisfy the continuity condition, at either the left end, $\psi(0) \neq 0$, or the right end $\psi(a) \neq 0$. Now, if we change A keeping B and n fixed, mainly the amplitude of the wave function changes, not the end values. So A acts like a scale factor, but it does nothing to meet the continuity condition.

Next, let us see what changing B does while keeping others constant. We see that the left-end point moves most, going up or down with increasing or decreasing B. When $B = 0$, it leads to $\psi(0) = 0$, half of the continuity condition.

Last, there is only n left to adjust. Keeping $B = 0$ and A fixed, increasing n values increase the number of oscillations as more waves emerge from the right wall. Whenever n happens to be an integer, though, the wave function vanishes at the right wall $\psi(a) = 0$. This is just the other half of the continuity condition needed. We see that the formation of discrete states.

From these explorations, we find that a physically allowed wave function satisfying the continuity condition Eq. (7.9) can be achieved only if $B = 0$ and $n = $ integer. Then the wave function takes the form

$$\psi(x) = A \sin(n\pi x/a), \quad n = 1, 2, 3 \ldots \tag{7.10}$$

From Eq. (7.7), we obtain the energy as $E_n = \hbar^2 \pi^2 n^2 / 2ma^2$, also discrete. Apparently, the continuity condition, or the boundary condition, leads to both the allowed wave function and allowed energy together (Exercise 7.6). The allowed energies and its associated wave function are known as eigenvalues and eigenfunctions (or eigenvectors). Together they form eigenstates (see Sects. 6.1 and 6.2.3).

Now what about A? As we have seen, it controls the amplitude of the wave function only. It may be determined by requiring the total probability of finding the particle to be 1, or $\int |\psi|^2 dx = 1$. The constant A is known as the normalization constant. We also see from Eq. (7.10) that $\psi(a) = A \sin(n\pi) = 0$ if n is an integer. That is why rewriting $k = n\pi/a$ is more convenient in this case.

We note that discrete eigenstates in quantum mechanics are a result of reasonable requirement for a well-behaved wave function. In fact, discreteness is not new or strange. We saw this in the discrete angles required to make a shot (Exercise 5.3). Similar requirement leads to discrete frequencies, or fundamental harmonics, in a string instrument. There, the boundary condition is that the string is fixed at two ends, and only discrete fundamental frequencies are allowed. The Schrödinger equation is a wave equation after all, and quantized energies are expected, naturally. We discuss quantum states more generally next.

7.3.3 Visualizing Quantum States

In the last section we have seen that in order to fit a known, general solution between the walls, only certain discrete energies can satisfy the box boundary conditions Eq. (7.9). The simplicity of the model shows clearly how it occurs naturally without any extraneous assumptions other than the wave function be continuous in the domain.

Simple solvable models like the particle in a box are few and far in between. In fact, the only other models that may be solved analytically in terms of elementary functions are the Dirac delta potential and the simple harmonic oscillator (SHO). Most realistic problems do not have closed-form solutions, and the boundaries can extend to infinity. Even so, the basic ideas demonstrated through the box boundary conditions will help us understand discrete quantum states in general potentials, via computation though.

Discrete energy levels exist in bound states where a particle's wave function is localized in a potential $V(x)$. The general boundary conditions are $\psi(-\infty) = \psi(\infty) = 0$, a straightforward extension of the box boundary conditions Eq. (7.9). An assumed but unstated property of the wave function is its smoothness, or the continuity of the first derivative. For well-behaved potentials, the Schrödinger equation Eq. (7.7) ensures a well-defined second derivative, mathematically requiring the first derivative to be continuous. Physically, the same is required for the particle flux to be continuous. Regardless, this fact will prove very useful in numerical computation next.

To solve Eq. (7.7) for a general potential, we discretize it over a space grid the same way the wave equation is solved (Sect. 6.2.2). Similar to Eqs. (4.4) and (6.19a), the second derivative is written as $d^2\psi/dx^2 = (\psi(x - \Delta x) - 2\psi(x) + \psi(x + \Delta x))/(\Delta x)^2$. Substituting it into Eq. (7.7) and rearranging it for convenience, we obtain

$$\psi(x + \Delta x) = \left(2 + \frac{2m(V(x) - E)}{\hbar^2}(\Delta x)^2\right)\psi(x) - \psi(x - \Delta x). \qquad (7.11)$$

Equation (7.11) is another three-term recursion relation in space, just like Eq. (6.21) in time. Likewise, we need two seed values to get it started as expected from a second-order differential equation. Guided by the boundary conditions $\psi(\pm\infty) = 0$, we set the domain in the range $[-R, R]$, assuming R is sufficiently large that the wave function is negligible beyond, that is $\psi(\pm R) \sim 0$. The range R depends on the characteristic length of the potential.

To ensure the boundary conditions are satisfied, our strategy is to start from the left $x = -R$, set $\psi(-R) = 0$ and $\psi(-R + \Delta x) = c_1$, and march upward. Similarly, we set $\psi(R) = 0$ and $\psi(R - \Delta x) = c_2$, and march downward. The nonzero constants $c_{1,2}$ are arbitrary because they affect only the overall normalization, and can be set to 1.

Now we pick a trial energy E, iterate Eq. (7.11) inward from both ends to some predetermined matching point x_m. When the upward and downward wave functions are matched up at x_m, they either connect to each other smoothly or not. If they do, we have found the right energy because the wave function satisfy the boundary conditions and is smooth at x_m. If they do not, we adjust the energy and repeat the process. This process of trial and error is known as the shooting method. As we have seen earlier, the way the wave function wiggles depends on the energy, and not every energy is expected to make the wave functions connect smoothly. Hence, the allowed energies will be discrete. Once the energy is found, so is the wave function, completing the eigenstate.

To determine the smoothness quantitatively, we can monitor the derivatives of the upward ($\psi_u'(x_m)$) and downward ($\psi_d'(x_m)$) iterations, and accept the solution if $|\psi_u'(x_m) - \psi_d'(x_m)|$ is sufficiently small.

Before actual computation, we need to choose a sensical units system because the SI units would be awkward and unnatural for quantum mechanical calculations. A suitable choice is the atomic units (a.u.) system, in which length, mass, and angular momentum are measured respectively in the Bohr radius (0.529 Å), the electron mass, and $\hbar$. In these units, the unit of energy is 27.2 eV.

If we assume the particle to be an electron, the practical advantage of the atomic units is that we can set $m = \hbar = 1$ in Eq. (7.11), so the recursion now reads simply

$$\psi_{n+1} = 2\left(1 + (V_n - E)(\Delta x)^2\right)\psi_n - \psi_{n-1}, \tag{7.12}$$

where $\psi_n \equiv \psi(x_n)$ and $V_n \equiv V(x_n)$ are the respective grid values at x_n as usual.

Numerically, Eq. (7.12) is readily applicable to any general potential, regardless of its shape. As the first example, we calculate the quantum states of the SHO with the potential $V(x) = \frac{1}{2}x^2$ in a.u. The exact eigenenergies for the SHO (Sect. 3.3) is $E_n = (n + \frac{1}{2})\hbar\omega$, $\omega = \sqrt{k/m}$. Here, the spring constant $k = 1$ and $m = 1$ in a.u., so $E_n = n + \frac{1}{2}$ a.u. The SHO is an important nontrivial problem with known analytic solutions, and is an ideal test case.

Computation is carried out with Program A.5.4 and sample results are shown in Fig. 7.5. The energy is selected with the slider. At each energy, the wave function is interactively visualized separately for the upward ψ_u and downward ψ_d iterations so the connection at the matching point may be seen more clearly. The relative difference between the first derivatives of the two parts is calculated as $\varepsilon = (\psi_d' - \psi_u')/\max(|\psi_d'|, |\psi_u'|)$ at x_m. It is basically a relative error with the sign information. For better accuracy, the three-point formula Eq. (4.4) is used for the derivatives $\psi_d'(x_m)$ and $\psi_u'(x_m)$. The derivations are almost a byproduct of the shooting method, and readily obtained from the wave functions.

An energy is allowed when the two parts smoothly join each other at the matching point. Practically, it is unlikely for the error ε to be zero exactly because of numerical error, so we accept the energy if the error is within some tolerance, $|\varepsilon| < 0.05$ in this case. As the trial energy increases from $E = 1.1$ to 1.5, both parts of the wave function start to get shallower, and the joint moves lower, eventually crossing the axis between $E = 1.2$ and 1.3 (mostly due to increased oscillatory behavior of the upward iteration. Run the program to see the interesting dynamic change and how the dashed part of the wave function flips from positive to negative due to matching). The connection at the joint becomes smoother, accompanied with progressively smaller error. At $E = 1.5$, the error $|\varepsilon| = 0.002$ is the smallest, and the energy is accepted. The exact energy for this SHO is $E_n = n + \frac{1}{2}$, so we just found the

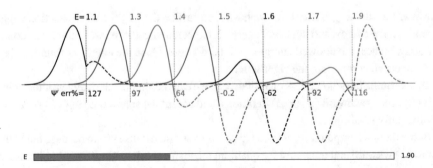

Fig. 7.5 The wave functions ψ of a simple harmonic oscillator ($V = \frac{1}{2}x^2$) at different trial energies (a.u.) labeled at the top. They are scaled so the maximum magnitude is 1, and also shifted to minimize overlap for clarity (see Program A.5.4 for dynamic change as energy is being selected from the slider). The solid curve in each wave function is the result of upward iteration and the dashed curve that of downward iteration. Both parts are matched at the dotted line (slightly off-center). The mismatch error is shown as the relative difference between the derivatives Ψ' of the two parts. The error is nearly zero when $E = 1.5$, which is the exact result $E_n = n + \frac{1}{2}$ for $n = 1$ in the chosen units (a.u.)

energy with $n = 1$, the first excited state of the SHO. Indeed, the single node (crossing zero) in the wave function confirms it also.

With the trial energy increasing still, the joint bends other way, increasing the error ε. Between the last two energies, the dashed part of wave function flips again due to matching, creating another node. It is well on its way to the second excited state at $E = 2.5$ ($n = 2$, two nodes). While these observations add insight to why discrete quantum states exist, nature (or the Schrödinger equation) does not care one way or the other. At a given energy, a state either exists or not.

Program A.5.4 performs well in other cases. In fact, it can be used as a robust general Schrödinger equation solver. We just need to replace the potential function and adjust other parameters as necessary. We leave it to the interested reader to explore other cases, including the finite square well, the particle in a box already discussed earlier, and even some 3D problem like the hydrogen atom (see Exercise 7.7).

7.4 Quantum Computing

From diffraction to double-slit interference, the principle of superposition is at the core of quantum mechanics. Superposition is a natively supported form of parallel processing in computing with quantum mechanical laws. Here we introduce the very basic elements of quantum computing, including basis states, superposition, qubits, and quantum gates. We put everything together with a quantum simulation of spin measurement and discussion of a quantum search algorithm.

7.4.1   Basis States and Principle of Superposition

As seen above, many systems can be characterized by discrete, stationary states. Given a system, the number of such states may be finite or infinite, though for our purpose we assume a finite number of states. These states are called basis states and they form a so-called basis set. We can think of the basis states as equivalent to the unit vectors in the Cartesian coordinates, except they represent physical states that can be selected via measurement.

For instance, electrons have spin that exists in two states (directions): up or down. The spin state can be selected by passing an electron through a magnetic-field selector (Stern-Gerlach apparatus, Fig. 7.6a), and the up and down states can be separated (see McIntyre (2012)).

An important property of quantum mechanics is that the system can exist in any state that is a linear combination of the basis states. This is known at the principle of superposition, a phenomenon we have seen in double-slit interference Sect. 7.2. With the spin as the example, let us denote the up and down states as $|\uparrow\rangle$ and $|\downarrow\rangle$, respectively ($|0\rangle$ and $|1\rangle$ are often used elsewhere (see Abhijith et al. (2022))). The vertical bars and brackets are part of compact Dirac notations that are convenient for algebraic operations. We use them here following convention, but for our purpose they could easily be replaced with just $\uparrow$ or $\downarrow$ without the bars and brackets). Then a possible state $|\psi\rangle$ of the electron spin is given by

$$|\psi\rangle = c_1|\uparrow\rangle + c_2|\downarrow\rangle \quad \text{(qubit)} \tag{7.13}$$

where c_1 and c_2 are constants (complex in general; also known as amplitudes) subject to $|c_1|^2 + |c_2|^2 = 1$. Using the vector analogy, we can think of superposition as equivalent to a vector composed of different components along the unit vectors.

The system can be prepared in the state $|\psi\rangle$ and will remain (evolve) just fine without external disturbance. If, however, we wish to measure the spin, we will get either $|\uparrow\rangle$ or $|\downarrow\rangle$, not both simultaneously (in the famous Schrödinger's cat game, we would see either a

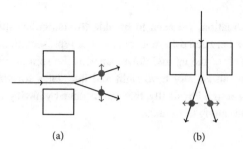

(a)   (b)

Fig. 7.6   Spin selector (Stern-Gerlach apparatus). (a) The incident electron passes through a magnetic-field selector in the z direction, separating the up and down states. (b) The magnetic-field selector is in the x direction, separating the right and left states

live cat or a dead cat). The probability of getting $| \uparrow \rangle$ is c_1^2, and that of $| \downarrow \rangle$ is c_2^2. We have no way of knowing the outcome beforehand, only the probability. Such a state of superposition is purely quantum mechanical, and there is no classical equivalence (Of course there is superposition of waves in classical physics, but there the waves do not represent discrete physical states as they do in quantum mechanics).

Suppose we obtained the $| \uparrow \rangle$ state in a measurement, the system will remain in the $| \uparrow \rangle$ state, and will not evolve back to a superposition like Eq. (7.13). In other words, measurement destroys (collapses) the superposition. But, as stated earlier, until measurement is carried out, the system remains in perfect state of superposition. In fact, Eq. (7.13) is not limited to two states, there can be N states ($N \gg 1$), and quantum mechanics can keep track of them simultaneously. In other words, parallel processing is a built-in feature of quantum mechanics. It is precisely this feature that enables quantum computing to be superior to digital computing for certain problems.

Focusing on two-state systems like the spin, Eq. (7.13) serves as the basic unit of quantum information, called a qubit. Unlike the 0 or 1 bit in digital computing, both $| \uparrow \rangle$ and $| \downarrow \rangle$ are represented (or co-exist) in a qubit, until measurement. For mathematical manipulations, it is convenient to represent the basis states by matrices

$$| \uparrow \rangle = \begin{pmatrix} 1 \\ 0 \end{pmatrix}, \quad | \downarrow \rangle = \begin{pmatrix} 0 \\ 1 \end{pmatrix}, \tag{7.14}$$

such that a qubit Eq. (7.13) can be expressed as

$$|\psi\rangle = c_1 \begin{pmatrix} 1 \\ 0 \end{pmatrix} + c_2 \begin{pmatrix} 0 \\ 1 \end{pmatrix} = \begin{pmatrix} c_1 \\ c_2 \end{pmatrix}. \tag{7.15}$$

A qubit can be in the basis state $| \uparrow \rangle$ if $c_1 = 1$, $c_2 = 0$, in basis state $| \downarrow \rangle$ if $c_1 = 0$, $c_2 = 1$, or anything in between.

7.4.2   Qubit Manipulation

To obtain useful information, we need to be able to manipulate qubits. This is done by transforming the states between different basis sets or representations. Take the spin states for example. If we think of the up and down states as the spin orientations in the vertical direction (z-axis), we can equally have right and left states for spin orientations in the horizontal direction (x-axis). Physically, this can be done by having the reference direction (e.g. the magnetic field) along the x-axis.

Alternative Basis States

Let us denote the right and left states by $|+\rangle$ and $|-\rangle$, respectively. Quantum mechanically, if we start with the $|+\rangle$ (or $|-\rangle$)) state, pass it through a magnetic-field selector in the z direction, it will be transformed into

$$|+\rangle = \frac{1}{\sqrt{2}}(|\uparrow\rangle + |\downarrow\rangle) = \frac{1}{\sqrt{2}}\begin{pmatrix} 1 \\ 1 \end{pmatrix}, \quad |-\rangle = \frac{1}{\sqrt{2}}(|\uparrow\rangle - |\downarrow\rangle) = \frac{1}{\sqrt{2}}\begin{pmatrix} 1 \\ -1 \end{pmatrix}. \quad (7.16)$$

Conversely, if we swap the right/left and up/down states, we obtain similar relations. For example, $|\uparrow\rangle = \frac{1}{\sqrt{2}}(|+\rangle + |-\rangle)$. The right/left states form an alternative basis set because the $|\uparrow\rangle$ and $|\downarrow\rangle$ states can be expanded in the $|+\rangle$ and $|-\rangle$ states.

Equation (7.16) shows that the right/left states are in fact superpositions of the up/down states. The implications are interesting: if we know the spin is in the right state $|+\rangle$ horizontally, and subsequently measure it in the z direction (vertically), we will get either spin up $(|\uparrow\rangle)$ or down $(|\downarrow\rangle)$, randomly, and with equal probability. Similarly, if the spin is in the up (or down) state, we will get either spin right or left after passing through the selector in the x direction, as depicted in Fig. 7.6b. There is no classical analog to this. Classically, a vector pointing up (z-axis) will have no x-components, vice versa. Quantum mechanically, we cannot know the z- and x-components simultaneously due to the uncertainty principle (Eq. (7.6)), so this "weird" feature is possible (technically we say that the x and z components do not commute).

Quantum Gates

The mathematical operation corresponding to Eq. (7.16) is called a Hadamard gate, expressed as a square matrix

$$H = \frac{1}{\sqrt{2}}\begin{pmatrix} 1 & 1 \\ 1 & -1 \end{pmatrix}. \quad (7.17)$$

The Hadamard matrix H is also its own inverse, so $H^2 = 1$.

With Eq. (7.14) and (7.16), we can verify using matrix multiplication rule that

$$H|\uparrow\rangle = |+\rangle, \quad H|\downarrow\rangle = |-\rangle, \quad H|+\rangle = |\uparrow\rangle, \quad H|-\rangle = |\downarrow\rangle. \quad (7.18)$$

Physically, the Hadamard gate acts like a state converter that turns one state into another. Because it is its own inverse, acting on a state twice returns it to the original state, provided no measurement has taken place in between.

Another useful gate is the flip gate X defined as

$$X = \begin{pmatrix} 0 & 1 \\ 1 & 0 \end{pmatrix}. \quad (7.19)$$

When applied to the qubit Eq. (7.15), it flips the rows

$$X\begin{pmatrix} c_1 \\ c_2 \end{pmatrix} = \begin{pmatrix} c_2 \\ c_1 \end{pmatrix}. \quad (7.20)$$

We note the special cases, $X|\uparrow\rangle = |\downarrow\rangle, X|\downarrow\rangle = |\uparrow\rangle$. Physically, it corresponds to a selector reversing its magnetic field direction. This operation has a digital equivalent, the NOT gate.

There are plenty other gates, of course, an interested reader should consult other references (see Exercise 7.8, and Abhijith et al. (2022)).

Measurement

Now, let us discuss the measurement of a spin or qubit to find out its state. Physically, we just place a detector behind the selector (Fig. 7.6) to determine if a spin up or down is detected. Mathematically, we do it through an operation called projection, again borrowing nomenclature from ordinary vector operations.

Suppose we want to extract the amplitude c_1 or c_2 from Eq. (7.15). We can do so with a matrix multiplication as

$$\begin{pmatrix} 1 & 0 \end{pmatrix}\begin{pmatrix} c_1 \\ c_2 \end{pmatrix} = c_1, \quad \begin{pmatrix} 0 & 1 \end{pmatrix}\begin{pmatrix} c_1 \\ c_2 \end{pmatrix} = c_2. \tag{7.21}$$

Note that the first matrix in each term is a row matrix, identical to the matrix representation of the up and down states Eq. (7.14), respectively, but turned sideways (transposed). In Dirac notation, the turning operation is denoted by swapping the bar and bracket so $\langle \uparrow | = \begin{pmatrix} 1 & 0 \end{pmatrix}$, such that $c_1 = \langle \uparrow | \psi \rangle, c_2 = \langle \downarrow | \psi \rangle$, and so on. Such operations are called projections, akin to the extraction of vector components in the Cartesian coordinates.

7.4.3 A Quantum Computer Simulator

We have now the very basic elements to explore several operations in quantum computing on qubits. Pending the arrival of a real quantum computer on everyone's laps, we will have to simulate the operations on the digital computer. However, the essential ideas are clearly demonstrated in the simulator nonetheless (see Program A.5.5).[3]

We aim to develop a simple quantum simulator to manipulate a qubit and do some simple measurements. See Sect. 7.4.4 on using a ready-made sympy qubit-manipulation library. First, we need to be able to create a qubit. Instead of obtaining it from an actual quantum computer, we can simulate this with a matrix $\begin{pmatrix} 1 \\ 0 \end{pmatrix}$, as given in Eq. (7.14).

Next, if we make a measurement on this qubit, we would get 1 (spin up) always. To make it a bit more interesting, let us apply the Hadamard gate to it. This will involve a matrix multiplication as shown in Eq. (7.18).

Third, we can simulate a readout measurement by projection according to Eq. (7.21). Putting it together, we have the following code to accomplish these functions.

---

[3] There are also other resources where one can learn about the basics of quantum computing and programming. For example, the Quantum Katas (see Microsoft (2022)) consist of a balanced range of tutorials and exercises using the Q-sharp (Q#) programming language.

```
In [1]:  def Qubit():            # returns spin up [1,0]
             return np.array([1, 0])

         def H(qubit):         # Hadamard gate
             hg = np.array([[1,1], [1,-1]])/np.sqrt(2.0)
             return np.dot(hg, qubit)

         def M(qubit):         # Measurement
             return np.dot([1, 0], qubit)
```

Here `Qubit()` returns a qubit as a 2×1 array. The function `H()` creates the Hadamard matrix represented by a 2×2 array. The Hadamard operation is performed via `np.dot()` which returns the result of the matrix multiplication between the Hadamard matrix and the input qubit. Finally `M()` carries out the projection of the input qubit onto the spin up state.

We now can simulate a number of computations. Let us compute the fraction of spin up states after a Hadamard gate. The process as implemented in Program A.5.5 is as follows: First, prepare a pure qubit $|\uparrow\rangle$ according to Eq. (7.14); then apply the Hadamard gate to it. Next, a "measurement" is performed to sample whether we get a spin up or down. The process is repeated many times so the fraction, or probability, can be meaningfully calculated from the ratio of the numbers of spin-up count to the total number of samplings. The code segment to do this is given below.

```
In [2]:  q = Qubit()
         hq = H(q)
         c1 = M(hq)          # amplitude
         r = random()
         if (r < c1**2): # spin up
             s = s + 1.0
```

Because we do not have a quantum gate to read out directly the spin state, the "measurement" is simulated by projecting the transformed state $|+\rangle$ on $|\uparrow\rangle$ to extract the amplitude c_1, then comparing the probability of getting a spin up c_1^2 with a random number. After repeated samplings, a statistically correct fraction can be determined. A sample output of Program A.5.5 gave the result of 0.502 with 1000 samplings. Different runs will usually produce different results within statistical fluctuations, but it should approach the exact value 0.5 with a larger number of samplings.

True Random Numbers and Schrödinger's Cat

The use of a digital random number generator is the least realistic part of the simulation, because it is only a pseudo-random number owing to the way random numbers are generated on the digital computer (see Sect. 8.1). The sequence of pseudo-random numbers closely resembles some statistical properties of being random, but is repeatable, hence is not truly random. In fact, only a quantum computer could generate a true random number due to the probabilistic nature of quantum measurement. To do so, we can associate a quantum readout of spin up and down with binary bits 1 and 0, respectively. Because a Hadamard-transformed state contains the two states with equal probability, the outcomes would be equally likely to be 1 or 0. We now have a random bit. To obtain a larger random number, we can generate as many bits as desired.

As the above simulation shows, a measurement of a Hadamard-transformed state yields either a spin up or down, not both. Repeating the measurement will tell us the probability of getting one state or the other (equal probability in this case). However we are unable to know which state will result just before the measurement: it is truly a random event. If we think of the two states as a cat being alive or dead respectively, we have a system that is a superposition of a cat in the live *and* dead states simultaneously, at least until measurement. If we wish to know by doing a measurement, it will reveal the cat being either alive *OR* dead. This illustrates the well-known Schrödinger's cat problem: measurement somehow alters the superposition (coherent state) irrecoverably: if you get a live cat, you will remain a live cat, and the same with a dead cat. Either will not become a superposition of being live *and* dead by itself. How does the act of measurement do that? Was the cat alive or dead immediately before the measurement?

Unfortunately, the question cannot be found satisfactorily from within quantum mechanics itself, because it, or more precisely the Schrödinger's equation, only describes how the state of superposition evolves in time. It does not say anything about measurement or a particular outcome, only the probabilities of the outcomes. This is the quantum measurement problem in a nutshell. Competing ideas (and schemes) of interpreting the measurement have been proposed. A commonly accepted viewpoint is the Copenhagen interpretation: The question is immaterial, and the system ceases to be a coherent superposition, which collapses to one state or the other when a measurement is made. In other words, measurement destroys the system. As unsatisfying as this answer is as to how the system stops being a superposition of states, it is an answer nonetheless. It does not affect the actual outcomes of predictions or measurements, only the interpretation. As stated earlier, it is in fact this very unique property of quantum mechanics keeping track of many states simultaneously, a kind of natural parallel processing, that makes quantum computing a powerful tool for suitable problems. One such problem is discussed in the next section.

7.4.4   Quantum Search: The Grover Algorithm

We now discuss a nontrivial but practical task where quantum computing excels: search. The problem is stated as follows: given an unordered list of $N \gg 1$ items, e.g., numbers, names, and so on, find the position of a target item as efficiently as possible. Classically, we go down the list, comparing each item with the target. The best case scenario occurs if the target item is at the beginning, and the worst case if it is at the end, so on average $N/2$ operations are required, or on the order of N. As it turns out, quantum computing can find the item with only $\sim \sqrt{N}$ operations, a quadratic speedup. It is possible with the Grover search algorithm discussed next.

7.4.4.1 Qubit Operations with Sympy

For this part of the discussion, it is convenient to use a readily available qubit manipulation library in `sympy` (see Sect. 3.3). Having developed a basic understanding earlier in Sect. 7.4, we can now focus on the algorithm without worrying about creating qubits, gates, and other basic operations. Still, let us familiarize ourselves with `sympy` qubit operations, starting with single-qubit manipulations. Run Program A.5.6 to follow along these operations interactively.

First we import the necessary library functions,

```
In [3]: from sympy import *
        from sympy.physics.quantum.qubit import Qubit, IntQubit, measure_partial
        from sympy.physics.quantum.qubit import measure_partial_oneshot
        from sympy.physics.quantum.gate import H
        from sympy.physics.quantum.qapply import qapply
        init_printing()
```

Both `Qubit` and `IntQubit` create qubits, the former taking as input a string of 0s and 1s, and the latter an integer. For instance, to create a single qubit, we do the following,

```
In [4]: q = Qubit(0)
        q
```

Out[4]: $|0\rangle$

Note the output is $|0\rangle$ instead of our usual $|\uparrow\rangle$. For single qubits, the following notations are equivalent, $|0\rangle \equiv |\uparrow\rangle, |1\rangle \equiv |\downarrow\rangle$.

A potential measurement of a qubit can be made with

```
In [5]: measure_partial(q,(0))
```

Out[5]: $(|0\rangle, 1)$

The `measure_partial` returns the result of a potential measurement of the qubit in the $|0\rangle$ state with a probability of 1 because that is the state the qubit was initialized to. It only predicts the probability if a measurement was carried out without actually making the measurement because doing so would collapse the qubit. To actually make a measurement, we use `measure_partial_oneshot`, which would reveal the outcome, not the probability of the outcome,

```
In [6]: measure_partial_oneshot(q, [0])
```

Out[6]: $|0\rangle$

Only one outcome is possible here. In actual measurement, the qubit would be destroyed.

To mix thing up a little, we can apply a Hadamard gate to the qubit to create a $|+\rangle$ state Eq. (7.18),

```
In [7]: hq = qapply(H(0)*q)
        hq
```

Out[7]: $\frac{1}{\sqrt{2}} (|0\rangle + |1\rangle)$

Here, H(0)*q says to apply the Hadamard gate to qubit 0 of the qubit. The result is generated only after qapply is called. It is a $|+\rangle$ state as expected from Eqs. (7.16) and (7.17).

A potential measurement of the $|+\rangle$ state yields

```
In  [8]:  measure_partial(hq,(0))
```

Out[8]: $(|0\rangle, \frac{1}{2}), (|1\rangle, \frac{1}{2}))$

We get both $|0\rangle$ and $|1\rangle$ with equal probability. If we conducted an actual measurement, we would get a single outcome,

```
In  [9]:  measure_partial_oneshot(hq, [0])
```

Out[9]: $|1\rangle$

We got $|1\rangle$, but could have gotten $|0\rangle$ with equal probability. To confirm, let us repeat the measurement 10 times,

```
In  [10]:  res = []
           for i in range(10):
               res.append(measure_partial_oneshot(hq, [0]))
           res
```

Out[10]: $[|0\rangle, |1\rangle, |1\rangle, |1\rangle, |0\rangle, |1\rangle, |0\rangle, |1\rangle, |0\rangle, |1\rangle]$

Out of 10 outcomes, we obtained $|0\rangle$ 4 times. If we repeated it another 10 times, we would get different outcomes. However, in a repeat measurement, we would have to do it with another qubit prepared in the same state because once a measurement is made, that state is destroyed. In this case, we would repeat the measurements on an ensemble of identically initialized states. Multiple qubits can be handled similarly, discussed in connection to the search method next.

7.4.4.2 Grover Algorithm

We need multiple qubits to handle more than two basis states. With n qubits, the number of such states is $N = 2^n$. In a two-qubit system, an example we will use to illustrate the search of a list, the possible states are $|00\rangle, |01\rangle, |10\rangle, |11\rangle$. For convenience, we shall label the states sequentially as $|\mathbf{0}\rangle, |\mathbf{1}\rangle, |\mathbf{2}\rangle, |\mathbf{3}\rangle$. They are also called ket states in a 4-dimensional vector space. For every ket state, there is a bra state, noted as $\langle\mathbf{0}|, \langle\mathbf{1}|$, and so on.

Any other state (state vector) can be written as a superposition of the ket states. A particularly useful state for our purpose is the uniformly weighted state,

$$|\psi_0\rangle = \frac{1}{2}\big[|\mathbf{0}\rangle + |\mathbf{1}\rangle + |\mathbf{2}\rangle + |\mathbf{3}\rangle\big]. \tag{7.22}$$

Analogous to Eq. (7.13), all four coefficients (amplitudes) are equal, $c_1 = c_2 = c_3 = c_4 = 1/2$, ensuring $|\psi_0\rangle$ is normalized, that is $\sum_s c_s^2 = 1$. The order of the ket states in Eq. (7.22) is unimportant quantum mechanically, though it may well be if they represent items in a list. For example, suppose the four ket states represent the letters ABCD sequentially. The list could be scrambled to BADC, a different list with letter positions changed, but it is still the same quantum state. The search method discussed next can still find the correct position of a given letter as long as they are associated with the ket state number.

Quantum Search: A Worked Example

Suppose we wish to find the position of letter B, called the target, associated with $|1\rangle$ in Eq. (7.22). The Grover search method works to boost the target state's amplitude, c_t, and simultaneously suppress the amplitudes of other states. When c_t is sufficiently dominant such that its probability c_t^2 is nearly 1, a measurement is made to reveal ket state, and the target position.

Let us now look at the Grover search method in action through a worked example and then discuss the general approach. Also see Fig. 7.7 for a graphical overview. We start with the uniform initial state Eq. (7.22). There are two steps involved.

First step. The amplitude of the target state is flipped but those of others are unchanged. In Eq. (7.22), we set $c_t \equiv c_2 \to -c_2$,

$$|\psi_1\rangle = \frac{1}{2}\big[|0\rangle - |1\rangle + |2\rangle + |3\rangle\big].\tag{7.23}$$

The normalization of $|\psi_1\rangle$ remains unchanged at 1.

Second step. All amplitudes are changed to $2\bar{c} - c_s$, where $\bar{c} = \sum_s c_s/4$ is the average amplitude. With $\bar{c} = [(1 - 1 + 1 + 1)/2]/4 = 1/4$, we have

$$|\psi_2\rangle = 0 \times |0\rangle + 1 \times |1\rangle + 0 \times |2\rangle + 0 \times |3\rangle.\tag{7.24}$$

The effect is that the target state amplitude (magnitude) is boosted to $c_t = c_2 = 1$ and all others are reduced to zero ($c_1 = c_3 = c_4 = 0$). A measurement (projection) will reveal only state $|1\rangle$, or B in the second position. The Grover search is successful.

Let us emphasize three points. First, the normalization of $|\psi_{1,2}\rangle$ is unaltered throughout as required between state to state transformations in quantum mechanics. Second, the amplitudes are effectively flipped *about* the average amplitude $\bar{c}$ after the operation $2\bar{c} - c_s$. To see this, rewrite $2\bar{c} - c_s = \bar{c} - d$ with $d = c_s - \bar{c}$ being the difference between c_s and $\bar{c}$. If d is positive (c_s above $\bar{c}$), then the result $\bar{c} - d$ flips d down from $\bar{c}$; otherwise it flips $|d|$ up. Third, the target state $|1\rangle$ could be in any other position at the start in Eq. (7.22), say at the first position, the only difference would be $c_t = c_1$ in the end. These same points are valid in the general approach discussed next.

Grover Method: General Approach

In a general quantum search problem, we would not know beforehand the position of the target state, noted as $|\mathbf{q_t}\rangle$, in the superposition; we only know its identity (qubit pattern, q_t). But this is sufficient to build a transformation to flip the target state in the first step. This magic can be achieved easier than one might have thought, at least conceptually.

At a simplified conceptual level, we seek a transformation $\hat{O}$ as a diagonal matrix. For the 2-qubit problem, it looks like

$$\hat{O} = \begin{bmatrix} (-1)^f & & & \\ & (-1)^f & 0 & \\ & 0 & (-1)^f & \\ & & & (-1)^f \end{bmatrix}.$$

(7.25)

Among other things, this form of matrix preserves the normalization (namely, it is unitary). For n qubits, the dimension of the $\hat{O}$ matrix would be $2^n \times 2^n$.

The function $f = f(q, q_t)$ is a boolean function that compares the input qubits q with the target qubits q_t,

$$f(q, q_t) = \begin{cases} 1 & \text{if } q = q_t \\ 0 & \text{otherwise} \end{cases}.$$

(7.26)

We note that the qubits in the function f are not actually measured in the classical sense for the comparison, because any measurement would collapse the state prematurely. It is part of the quantum gate logic at the hardware level.

Now imagine we iterate through the ket states in Eq. (7.22) to build $\hat{O}$, assuming the second state $|1\rangle$ as the target, f will be 0 for all but the second diagonal, and we will obtain

$$\hat{O} = \begin{bmatrix} 1 & & & \\ & -1 & 0 & \\ & 0 & 1 & \\ & & & 1 \end{bmatrix}.$$

(7.27)

Only the second diagonal will be -1. We can verify with matrix multiplication that given a general state $|\psi\rangle = [c_1, c_2, c_3, c_4]^T$, the result of $\hat{O}|\psi\rangle$ is

$$|\psi'\rangle = \hat{O}|\psi\rangle = \begin{bmatrix} 1 & & & \\ & -1 & 0 & \\ & 0 & 1 & \\ & & & 1 \end{bmatrix} \begin{bmatrix} c_1 \\ c_2 \\ c_3 \\ c_4 \end{bmatrix} = \begin{bmatrix} c_1 \\ -c_2 \\ c_3 \\ c_4 \end{bmatrix},$$

(7.28)

where only the target amplitude is flipped.

The transformation $\hat{O}$ will always flip the correct amplitude corresponding to the position of the target state. The matrix $\hat{O}$ Eq. (7.25) is also known as the oracle gate. Once it is known, the first step in the Grover method is solved, and the resultant is a new state

$|\psi'\rangle = [c_1', c_2', c_3', c_4']^T$ which can be regarded as a reflection in a two-dimensional Cartesian space.[4]

In the second step, the amplitude boosting operation is carried out as

$$|\psi''\rangle = \hat{W}|\psi'\rangle = \begin{bmatrix} 2\bar{c} - c_1' \\ 2\bar{c} - c_2' \\ 2\bar{c} - c_3' \\ 2\bar{c} - c_4' \end{bmatrix}, \quad \hat{W} = (2\bar{c} - 1)\mathbf{1}, \quad \bar{c} = \frac{1}{N}\sum_s c_s'. \tag{7.29}$$

Because this operation amplifies the target amplitude and diminishes others, it is known as the winner gate, or $\hat{W}$ gate. Interestingly, it also has a geometric interpretation as a reflection about the initial state.[5]

Together with the $\hat{O}$ transformation, the $\hat{W}$ transformation will increase the target state amplitude, and reduce the amplitudes of other states (see Fig. 7.7). Finally, here is the Grover algorithm:

1. Initialize the uniform state $|\psi_0\rangle$, Eq. (7.22), and build the oracle gate $\hat{O}$, Eq. (7.25).
2. Flip the target amplitude with $|\psi'\rangle = \hat{O}|\psi_0\rangle$.
3. Boost the target amplitude with $|\psi''\rangle = W|\psi'\rangle$.
4. Replace $|\psi_0\rangle$ with $|\psi''\rangle$ and repeat step 2.

As explained in the footnotes, the iterations may be regarded as a series of successive reflections, effectively rotating the initial state onto the target state. It can be shown that the

---

[4] The $\hat{O}$ gate has a useful geometric interpretation. To see this, let us separate the target state $|t\rangle$ from the rest, $|\psi'\rangle = \hat{O}|\psi\rangle = \sum_{s\neq t} c_s|s\rangle - c_t|t\rangle = |\psi_\perp\rangle - c_t|t\rangle$, where $|\psi_\perp\rangle = \sum_{s\neq t} c_s|s\rangle$, the original state minus the target state. Incidentally, $|\psi_\perp\rangle$ is perpendicular to the target state because $\langle\psi_\perp|t\rangle = \sum_{s\neq t} c_s\langle s|t\rangle = 0$ due to the basis states being orthogonal to each other. Thus, we can regard $|\psi_\perp\rangle$ and $|t\rangle$ as the $\hat{x}$ and $\hat{y}$ directions, respectively, in the two-dimensional Cartesian coordinates. Before the transformation, we can visualize the state $|\psi\rangle$ as a vector slightly above the $\hat{x}$ axis. Because the transformation flips the $\hat{y}$ component of $|\psi\rangle$ and leaves the $\hat{x}$ component $|\psi_\perp\rangle$ unchanged, the transformed state $|\psi'\rangle$ can be visualized as a reflection of $|\psi\rangle$ about the $\hat{x}$ axis, so it is symmetrically below the axis. Furthermore, the direction of $|\psi_\perp\rangle$ remains unchanged in a subsequent transformation in the second step (W gate, see discussion following Eq. (7.29)), so $|\psi_\perp'\rangle$ points to the fixed $\hat{x}$ direction throughout. Its magnitude get progressively reduced, however.

[5] The $\hat{W}$ gate is equivalently defined as $\hat{W} = 2|\psi_0\rangle\langle\psi_0| - 1$, where $|\psi_0\rangle = [1, 1, \ldots]^T/\sqrt{N}$ is the uniform state Eq. (7.22). It follows that $\langle\psi_0|\psi\rangle = \sum_s c_s/\sqrt{N} = \bar{c}\sqrt{N}$, and $\hat{W}|\psi\rangle = 2\bar{c}[1, 1, \ldots]^T - |\psi\rangle$, same as Eq. (7.29). Like the $\hat{O}$ gate in the first step, the W gate is unitary and also has a geometric interpretation. Let $|\psi'\rangle = \hat{W}|\psi\rangle$. The operation is unitary, because $\langle\psi'|\psi'\rangle = \langle\psi|\psi\rangle$. The projection on $|\psi_0\rangle$ is unchanged, $\langle\psi_0|\psi'\rangle = 2\langle\psi_0|\psi_0\rangle\langle\psi_0|\psi\rangle - \langle\psi_0|\psi\rangle = \langle\psi_0|\psi\rangle$, so the parallel component remains the same. If there is a perpendicular component prior to the operation, say $|\psi_p\rangle$, then afterwards, $|\psi_p'\rangle = W|\psi_p\rangle = 2|\psi_0\rangle\langle\psi_0|\psi_p\rangle - |\psi_p\rangle = -|\psi_p\rangle$ because $\langle\psi_0|\psi_p\rangle = 0$ by definition. So the perpendicular component is reversed. Therefore, the $\hat{W}$ gate is a reflection about the uniform state vector $|\psi_0\rangle$. Together, the $\hat{O}$ and $\hat{W}$ transformations are a reflection about the $\hat{x}$ axis followed by another reflection about the initial state $|\psi_0\rangle$.

target amplitude grows until $\pi \sqrt{N}/4$ iterations have been reached. The target amplitude c_t will be very close to 1 for large N (it is exactly 1 only of 2-qubit systems), and it will shrink after that point. This is the optimal time to make a measurement to find out the target state (position in the list).

To simulate the process on the digital computer, we would expect computational cost to be about $N\sqrt{N}$, because the $\hat{O}$ and $\hat{W}$ gates would need N operations in each iteration, so the Grover method offers no advantage on a digital computer, no surprise here (see Exercise 7.11). But a quantum computer keeps track of all the qubits in parallel, and activates the gates simultaneously, so the computational cost is $\propto \sqrt{N}$. The speedup can be considerable. Suppose a given search takes one second on both a digital and a quantum computer. If the list is 100 times larger, it will take 100 s on the digital computer, but only 10 s on the quantum computer.

Grover Search with Sympy

Whereas the Grover iterations for the minimum two-qubit system is easy to work out by hand, three or more qubits are not as easy. However, sympy (Sect. 3.3) provides a convenient platform on which we could test drive a Grover search engine as if we were programming on an actual quantum computer. We illustrate the method here with three qubits for clarity, but more qubits could be used. We recommend the reader to run interactively Program A.5.6 and also refer to Fig. 7.7 about the process graphically.

First we import the necessary Grover modules from sympy,

```
In [11]:  from sympy.physics.quantum.grover import OracleGate, WGate
          from sympy.physics.quantum.grover import grover_iteration, apply_grover
```

The OracleGate and WGate modules correspond to our transformations $\hat{O}$ Eq. (7.25) and $\hat{W}$ Eq. (7.29), respectively. The grover_iteration completes one Grover iteration (first and second steps), and apply_grover solves the problem complete as seen shortly.

Next we set the number of qubits and construct the $\hat{O}$ and $\hat{W}$ transformations,

```
In [12]:  n = 3    # num of qubits
          f = lambda basis: basis == IntQubit(Qubit('001'))
          O = OracleGate(n, f)   # Oracle operator
          W = WGate(n)
```

As discussed earlier, a truth function f Eq. (7.26) is necessary for building $\hat{O}$. We choose the target state to be '001', or the second ket state $|1\rangle$. We use Qubit('001') to ensure it is a three-qubit system and convert to an integer qubit with InQubit because the latter alone would be ambiguous as to how many qubits are used. The function compares the input (basis) with the target state, and returns true (1) if they match, and false (0) otherwise. The function f is passed to OracleGate along with n, the number of qubits. The WGate function needs only the number of qubits.

We next initialize the uniform state $|\psi_0\rangle$ Eq. (7.22),

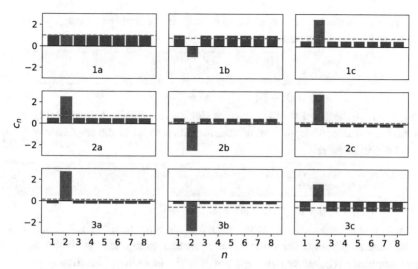

Fig. 7.7 Amplitudes in Grover iterations of a 3-qubit system. The amplitudes are scaled by $2^{-3/2}$ (initial normalization factor) for tidiness. Each row represents one iteration. The red color (second bar) indicates the target state. The dashed line marks the average amplitude in each configuration. In the first iteration (top row), the starting state (1a) is the uniform state Eq. (7.22). The first step (1b) flips the target state Eq. (7.28), and the second step (1c) boosts the amplitude of the target state and reduces the rest Eq. (7.29). The last configuration (1c) serves as the starting state of the second iteration (2a). The optimal Grover iteration stops at (2c) where the amplitudes of the target and non-target states are $11/8\sqrt{2}$ and $-1/8\sqrt{2}$, respectively, yielding a ratio of probabilities to 121/1. The ratio would become progressively larger with more qubits, so the probability of the target state being revealed in a measurement is nearly certain (see text). The third iteration is shown for comparison purpose only, and the target state becomes less dominant at the end (3c). Further iterations will eventually lead back to the starting point (1a) and the cycles repeats

```
In [13]:  c = sqrt(2)/4  # cancel factor
          qs = IntQubit(Qubit('000'))
          psi0 = H(2)*H(1)*H(0)*qs
          psi0 = qapply(psi0)
          simplify(psi0/c)
```

Out[13]: $|0\rangle + |1\rangle + |2\rangle + |3\rangle + |4\rangle + |5\rangle + |6\rangle + |7\rangle$

Starting with three qubits '000' ($|\uparrow\uparrow\uparrow\rangle$), three successive Hadamard transformations Eq. (7.18) are applied to each qubit respectively. Because a single Hadamard operation H(i) turns qubit i into an equal superposition of two states (up and down) Eq. (7.16), three operations in a row produce a superposition of eight ket states with equal amplitudes (Fig. 7.7, 1a) like Eq. (7.22) but with eight terms (also called outer product). Operations are queued and actuated with qapply. The amplitudes are scaled by the overall normalization factor $1/\sqrt{8}$ to make the output tidy (actual computations use unmodified amplitudes, of course).

We can peek in the Grover method step by step. In the first step, we apply the $\hat{O}$ transformation to flip the target state,

```
In [14]: psi1 = qapply(O*psi0)   # 1st step, reverse target amplitude
         simplify(psi1/c)
```

Out[14]: $|0\rangle - |1\rangle + |2\rangle + |3\rangle + |4\rangle + |5\rangle + |6\rangle + |7\rangle$

The amplitude of the target state $|1\rangle$ changes sign and the rest is unchanged (Fig. 7.7, 1b).

In the second step, we apply the $\hat{W}$ transformation to boost the amplitude of the target state and suppress the rest,

```
In [15]: psi2 = qapply(W*psi1)   # 2st step, apply W gate
         simplify(psi2/c)
```

Out[15]: $\frac{1}{2}(|0\rangle + 5|1\rangle + |2\rangle + |3\rangle + |4\rangle + |5\rangle + |6\rangle + |7\rangle)$

The target state is amplified five times relative to the other states (Fig. 7.7, 1c). The latter have the same amplitudes. The operation is equivalent to flipping the amplitude about the average amplitude. The above two steps complete the first Grover iteration.

The last state of the first iteration (Fig. 7.7, 1c) serves as the initial state of the second Grover iteration, now with nonuniform amplitudes. Repeating the two steps, we obtain

```
In [16]: psi1 = qapply(O*psi2)   # repeat 1st step
         simplify(psi1/c)
```

Out[16]: $\frac{1}{2}(|0\rangle - 5|1\rangle + |2\rangle + |3\rangle + |4\rangle + |5\rangle + |6\rangle + |7\rangle)$

```
In [17]: psi2 = qapply(W*psi1)   # repeat 2nd step, apply W gate
         simplify(psi2/c)
```

Out[17]: $\frac{1}{4}(-|0\rangle + 11|1\rangle - |2\rangle - |3\rangle - |4\rangle - |5\rangle - |6\rangle - |7\rangle)$

The intermediate state again shows the sign change of the target state (Fig. 7.7, 2b). The amplitude of the target state is increased also (Fig. 7.7, 2c).

The amplitude of the target state reaches its maximum value at iteration $[\pi\sqrt{2^n}] = 8/4 = 2$ for a three-qubit search, so we should stop the iterations. The amplitudes of the target and non-target states are $11/8\sqrt{2}$ and $-1/8\sqrt{2}$, respectively, yielding a ratio of $r = -11/1$ after the second iteration. The ratio of probabilities is $r^2 = 121/1$. This relative probability is large but not overwhelming. If a measurement is made, it would reveal the target state with a probability ~ 0.945 (Exercise 7.10). Recall for a two-qubit search it is 1, which happens to be accidental.

If we iterate one more time, we see that (Fig. 7.7, 3a–3c) the target amplitude actually decreases, and if we keep on going, it would reach the uniform state and the cycle repeats.

We can combine the two steps of an iteration and skip the intermediate results,

```
In [18]: psi1 = qapply(grover_iteration(psi0, O))   # one iteration
         simplify(psi1/c)
```

Out[18]: $\frac{1}{2}(|0\rangle + 5|1\rangle + |2\rangle + |3\rangle + |4\rangle + |5\rangle + |6\rangle + |7\rangle)$

```
In [19]: psi2 = qapply(grover_iteration(psi1, 0))   # another one
         simplify(psi2/c)
```

Out[19]: $\frac{1}{4}(-|0\rangle + 11|1\rangle - |2\rangle - |3\rangle - |4\rangle - |5\rangle - |6\rangle - |7\rangle)$

The results are as expected.

Finally, we can even combine all necessary iterations and solve the problem automatically with `apply_grover`,

```
In [20]: sol = qapply(apply_grover(f, n))   # fully automatic
         simplify(sol/c)
```

Out[20]: $\frac{1}{4}(-|0\rangle + 11|1\rangle - |2\rangle - |3\rangle - |4\rangle - |5\rangle - |6\rangle - |7\rangle)$

It relieves us of having to build the transformations manually. We just need to tell the solver what the target state is via the f function. It stops at the optimal iteration and gives the final results.

Lastly, let us examine the rate of convergence of the Grover method with the number of qubits n. As seen earlier, the Grover method gets better with each iteration, reaching the highest probability p_n up to the optimal iteration, then gets worse past that. This is in contrast to most other methods that get more accurate with more iterations. Interestingly, the probability approaches 1 nonmonotonically. For instance, the optimal probabilities for 4 to 6 qubits searches are respectively, 4: 0.96131, 5: 0.99918, and 6: 0.99658.

Of course, practical gain within quantum search is significant only for large n. Let ϵ_n be the error of the probability p_n from unity, namely $\epsilon_n = 1 - p_n$. For larger n, the error scales roughly as $\epsilon_n \simeq 2^{-n}$. So on average, for every 3.3 qubits, we obtain an additional digit of accuracy. Just for the fun of it, with 40 qubits (list size 1,099,511,627,776), the optimal converging number of iterations is 823,549, and the probability to 15 significant digits (limit of double precision) is 0.999999999999592, for an error of $\epsilon_{40} \sim 10^{-13}$ (see Exercise 7.11). The probability of finding the target state is practically a certainty. To put it in perspective, imagine we have been watching a basketball game by two evenly-matched teams that have been playing continuously since the early modern human era (300,000 years ago, day and night, some athletes!), and we take snapshots of the basketball court once every second. A probability of 10^{-13} would be equivalent to seeing that all but *one* of these snapshots would show the ball in the same half of the court. An extremely unlikely event indeed.

7.5 Exercises

7.1. Relative perspective. Relativity is all about observing and relating events from different perspectives but using the same laws and definitions. In Sect. 7.1.2 we discussed time and space from Stacy's perspective (Fig. 7.1). Here, we explore it from Moe's perspective.

(a) Graph Stacy's position x' as a function of time t' in Moe's reference frame (the train). Assume $\beta = 0.5$. Is it different from what you would expect nonrelativistically? Explain.

(b) Graph the times Δt vs. $\Delta t'$ for different β values, e.g., $\beta = 0, 0.2, 0.5, 0.99$. Visualize the graph using the slider widget and template Program A.5.1.

7.2. A thought experiment: Laser tagging. Due to the unattainable high speed needed for relativistic effects to be noticeable under normal circumstances, we often rely on thought experiment to examine relativistic scenarios. We consider such a scenario involving Stacy on the platform and Moe on the train moving with velocity v mentioned in Sect. 7.1.1.

(a) Let us assume at $t = 0$ Stacy fires a laser pulse toward a pillar at a distance d away. The (outgoing) laser pulse hits the pillar after $t_1 = d/c$, and is reflected back to her after $t_2 = d/c$. So naturally the total round trip time is $t = t_1 + t_2 = 2d/c$. Work out the corresponding times t_1' and t_2' from Moe's perspective. Keep in mind that (i) the speed of light is the same; (ii) during the outgoing trip the pillar is moving toward the pulse with $-v$, and during the return trip Stacy is moving away from the pulse with $-v$; and (iii) the pillar's position is contracted in Moe's reference frame. Compare Moe's primed times with Stacy's unprimed ones. Which two constitute a time-dilation pair?

*(b) Graph the positions of the laser pulse, Stacy, and the pillar as a function of time in Moe's reference frame for a given β value. Express distance and time in units of d and d/c, respectively. Explain the three curves.

Once the graph looks correct, make an interactive plot with a slider for choosing the β value. Comment on the graphs for small $\beta < 0.1$ and for large $\beta > 0.9$.

7.3. Single-slit diffraction.

(a) Investigate single-slit diffraction with Program A.5.2. Make sure the radio button is on single slit. Observe how the diffraction patterns change when you change the wavelength λ or the width w while keeping the other one fixed. Describe your observations. For a given w, choose a wavelength λ such that at least two bright and two dark spots (maximum and minimum intensities, respectively) can be observed on either side of center peak. Record the x positions either visually by moving the mouse pointer to the right position and read off the value from the lower-left corner of the plot, or more accurately with automatic searching (see Sect. 4.5 and Program A.2.2). Next, do the same but for a chosen width w.

(b) The bright and dark spots are given by the following formulas,

$$\sin\theta_{min} = m\frac{\lambda}{w}, \quad \sin\theta_{max} = (m + \frac{1}{2})\frac{\lambda}{w}, \quad m = 1, 2, 3, \ldots \tag{7.30}$$

Convert the positions data recorded from part 7.3a earlier into angles, knowing the distance of the screen h (1 m, but check the program to make sure). Compare them with theoretical values in Eq. (7.30), and discuss any differences. Optionally, derive the relation for minimum intensities in Eq. (7.30).

7.4. Double-slit interference.

(a) Let n be the number of peaks in double-slit interference. Predict how n will change if you: (i) increase the separation d while keeping the wavelength λ and slit width w constant; and (ii) decrease w while keeping λ and d constant. Test your predictions with Program A.5.2. Explain the results. Also explain how the intensities change as w gets increasingly small.

(b) Quantify the number of peaks n in double-slit interference as a function of slit separation d. Vary d and count the number of peaks within the central diffraction profile. You should modify Program A.5.2 for automatic peak counting using the method of Program A.2.2. Once you have five or more data points, plot n vs. d. Is the trend linear? Explain.

(c) Find the conditions for minimum and maximum intensities (similar to Eq. (7.30) in Exercise 7.3) in terms of λ, d, and θ (Fig. 7.2) for double-slit interference. Check that they are consistent with data above.

*(d) Interference occurs when the sources are coherent (same wavelength and same phase difference). Simulate what happens when incoherent sources are involved. To so this, assume the same wavelength but add a random phase to the waves from the second slit. Sum the intensities over N trials, and average them. The pseudo code is as follows,

```
calc psi1 and psi2 from slit 1,2
set N trials
intensity = 0
for each trial:
    set random phase = 2*pi*random()
    add phase to psi2, eg e^(i phase)*psi2
    add |psi1 + psi2|^2 to intensity

avg. intensity = intensity/N
```

Integrate the code into `updatefig` in Program A.5.2. Try different values of $N = 10, 20, 100$, and so on. Is interference still prominent? Explain the results.

7.5. Uncertainty principle. Assume the incident wave in single-slit diffraction is a matter wave consisting of particles with vertical momentum p. First, calculate the horizontal momentum p_x for diffraction to the first minimum angle (Eq. (7.30)). The uncertainty in momentum is twice that, $\Delta p_x = 2|p_x|$. Next, using Eq. (7.5), show that the product of $\Delta x \Delta p_x$ agrees with Eq. (7.6) with $\Delta x = w$.

7.6. Box boundary condition.

(a) Analytically apply the box boundary condition Eq. (7.9) to the general solution Eq. (7.8) and obtain the solutions Eq. (7.10).

(b) In Sect. 7.3.1, we chose the origin at the left wall. Because the choice of a coordinate system should not affect the physics, let us now set the origin at the center of the well, so the range is $-\frac{1}{2}a \le x \le \frac{1}{2}a$. Instead of Eq. (7.9), what are the new boundary conditions? Modify Program A.5.3 to visualize the quantum states accordingly. Determine the allowed energies and wave functions. Explain the differences from the old coordinate system.

7.7. Visual quantum eigenstates. Visually and directly investigate quantum eigenstates for select potentials of varying complexity. The template program is Program A.5.4. Use atomic units throughout.

(a) We start with SHO discussed in text. What is the expected eigenenergies if the spring constant is doubled, $k = 2$? Now run Program A.5.4 with the potential $V = x^2$. Find at least three allowed energies, and compare with the expected values.

(b) Let us revisit another exactly known system, the infinite potential well. It is easier to work with a symmetric potential well extending from $-a/2$ to $a/2$, so restrict x to $-\frac{1}{2}a \le x \le \frac{1}{2}a$ where $V = 0$. Let $a = 2$, so $R = a/2 = 1$, and set $E_{max} = 12$. Run the program to find the allowed energies. Make sure the energy search range is adjusted correctly. How many states have you found? Compare their values with the exact results. If you double $a = 4$, how many states do you predict in the same energy range? Test your predictions.

(c) Find the eigenenergies in the finite square potential. Use the same parameters as that in Sect. 4.4, and compare results found there with the root solver method and here with wave function matching method. Do the same for the odd states discussed in Exercise 4.2. Compare and comment on the advantages and disadvantages of the two methods.

*(d) Now, let us consider the Morse potential,

$$V = V_0 e^{\alpha(1-x)}(e^{\alpha(1-x)} - 2). \qquad (7.31)$$

The range of x is $x > 0$, and we may regard $V(x \le 0) = \infty$, a hard wall at the origin. The Morse potential is useful for modeling interactions between atoms in molecules in the radial direction. For this exercise, choose the parameters as $V_0 = 1$, $\alpha = \sqrt{2}$. First, sketch the Morse potential by hand, then graph it on the computer. From the shape of the potential, what energy search range do you expect to find bound states? Next, extend Program A.5.4 to take into consideration of the asymmetric range and the corresponding boundary condition, $\psi(0) = \psi(\infty) = 0$. Find as many eigenenergies as possible, and graph the energy level diagram to scale. Describe the results. Now examine the low-lying states (the lowest 4 or 5 states) more carefully. Does the energy spacing between these states look similar to a certain system? Explain.

*(e) Finally, compute the energy levels of the hydrogen atom with the Coulomb potential, $V = -1/x$, $x \geq 0$. Use the extended code from part 7.7d if already developed. Take care to manage the singularity of the potential at $x = 0$ (you do not actually need $V(0)$ when you know $\psi(0) = 0$). Also, this potential decays slowly, so use a larger upper range R than normal. Increase the number of grid points accordingly to maintain accuracy. Again, find as many allowed states as you can. The energies calculated this way should in fact be the energies of the so-called s-states of hydrogen. Compare your results with the exact s-state energies, $E_n = -\frac{1}{2n^2}$, of the hydrogen atom.[7] Optionally, if your modified program works well, try variations of the potential, say $V = -1/x^{1+\delta}$ or $V = -\exp(-\delta x)/x$ for some small $\delta \ll 1$. Comment on your results in contrast to $\delta = 0$.

7.8. Qubits and gates.

(a) Verify that the left and right qubit states of Eq. (7.16) are normalized and orthogonal to each other.

(b) The phase gate is defined as,

$$R(\phi) = \begin{pmatrix} 1 & 0 \\ 0 & e^{i\phi} \end{pmatrix}. \tag{7.32}$$

It can be used to add a phase to the down state. Show that (i) R is unitary; and (ii) when applied to Eq. (7.15), we pick up a phase as

$$R\begin{pmatrix} c_1 \\ c_2 \end{pmatrix} = \begin{pmatrix} c_1 \\ e^{i\phi}c_2 \end{pmatrix}.$$

In particular, note that $R|\uparrow\rangle = |\uparrow\rangle$, and $R|\downarrow\rangle = e^{i\phi}|\downarrow\rangle$.

(c) Like many other gates, the Hadamard gate H Eq. (7.17) can be interpreted in terms of geometric operations such as rotation, reflection, and so on. Assume $|\uparrow\rangle$ and

---

[7] Due to a quirk of the Coulomb potential in 3D, the s-states energies are the same for non s-states as well because of degeneracy in angular momentum.

$| \downarrow \rangle$ states act as the respective $\hat{\mathbf{x}}$ and $\hat{\mathbf{y}}$ unit vectors in the Cartesian coordinates. Show that $H| \uparrow \rangle$ is equivalent to a rotation of $| \uparrow \rangle$ by 45° counterclockwise about the origin. Sketch the operation. What geometric operation(s) is $H| \downarrow \rangle$ equivalent to? Explain with a schematic sketch.

7.9. Quantum computer simulation. We will simulate spin measurement and generation of random bits in this exercise.

(a) Compute the average spin with the simulator Program A.5.5. Run it a few times, record the values. Adjust N if you wish.

Now, write a code to calculate it more authentically with the sympy simulator (Sect. 7.4.4.1). The code should be pretty similar to Program A.5.5, and the pseudo code for the main loop is

```
for each trial:
    create a qubit q
    apply Hadamard gate qapply(H(0)*q)
    measure the transformed qubit with measure_partial_oneshot
    add spin to sum if the outcome is the up state |0>

avg. spin = sum/N
```

(b) The average spin calculation process generates a random series of single qubits. Develop a code to generate a random series of double qubits with sympy such as $|00\rangle, |10\rangle, \ldots$. You can create two single qubits and manipulate and measure each independently, or create a double qubit and apply successive Hadamard gate to each (forming an outer product, see Sect. 7.4.4.2). What is the probability of getting two down spins, namely, $|11\rangle$? Compare your calculation with the theoretical value.

7.10. Quantum search. It is the nature of quantum mechanics that nearly everything is stated in terms of probabilities. Quantum search is no exception. We investigate amplitudes and probabilities in Grover search next.

(a) In a 3-qubit Grover search, the optimal number of iterations is 2 (Fig. 7.7 (2c)). Carry out one more iteration manually and find the amplitudes shown in Fig. 7.7 (3c). Check to make sure the state remains normalized.

(b) The probability of finding a non-target state, namely a false positive, is not negligible in Grover search if the number of qubits is small. To reduce the chance of a false positive, we might repeat the search again. Referring to the amplitudes in a 3-qubit Grover search (Fig. 7.7 (2c)), calculate the probabilities in two successive searches of finding: the target state twice; the target state and non-target state, each once; and the non-target state twice. Another possibility of reducing the likelihood of false positives would be to pad the search list to increase the number of qubits artificially. See next part.

(c) Consider a Grover search with $n = 4$ qubits. How many items can be handled? What is the expected optimal number of iterations? Conduct a Grover search. Give the final ratio of the amplitudes of the target state to non-target states, and the probability. Repeat the process for 5 and 6 qubits, and verify the probabilities given in the text. Push the search to 8 qubits or as high as you can depending on your computer speed, because the simulation becomes very slow for larger n (see Exercise 7.11). Calculate the error of the final probability from unity. Discuss how closely it follows the 2^{-n} scaling.

*7.11. Convergence rate of Grover search. As stated in the text, simulating Grover search on the digital computer offers no speed advantage. Sympy is no exception as it becomes increasingly slow with large number of qubits $n > 10$. But if we are only interested in the amplitudes and probabilities without actually performing the search (not bothering with building and applying the $\hat{O}$ and $\hat{W}$ transformations), then it can be done straightforwardly.

The basic steps of the Grover algorithm still apply but in a much simplified way. Here is a possible pseudo code, assuming the amplitudes of the target and non-target states are a and b, respectively.

```
def groveramp(n):
    let N = 2**n
    let m = int(pi*sqrt(N)/4), optimal iteration number
    initialize uniform amplitudes a=b=1/sqrt(N)
    for iterations up to m:
        flip a
        calculate avg amplitude: avg = (a + b*(N−1))/N
        boost a: a = 2*avg − a,  b = 2*avg − b
    return a, b
```

Write a program to calculate the amplitudes a, b, and the probabilities a^2 for qubits $n = 10$ to 40. Check that normalization is maintained at each n.

Plot the error of the probabilities from unity ϵ_n as a function of n. From the slope, find the average number of qubits needed for each digit of improved accuracy.

References

J. Abhijith, A. Adedoyin, J. Ambrosiano, P. Anisimov, W. Casper, and et al. Quantum algorithm implementations for beginners. *ACM Transactions on Quantum Computing*, pages 1–95, 2022. URL https://doi.org/10.1145/3517340.

D. J. Griffiths and D. F. Schroeter. *Introduction to quantum mechanics*. (Cambridge University Press, Cambridge), 3rd edition, 2018.

D. McIntyre. *Quantum mechanics: A paradigms approach*. (Pearson, Upper Saddle River, NJ), 2012.
Microsoft. Quantum Katas. *Microsoft Quantum Katas Project*, 2022. URL https://github.com/
    Microsoft/QuantumKatas.

Statistical and Thermal Processes

8

Many processes are inherently random rather than deterministic, such as the throw of dice or the motion of a dust particle in the air. Even so, there are often rules or statistical properties that arise out of randomness. Thermal processes are a prime source of this phenomenon. Seemingly random collisions of atoms and molecules producing a net heat flow is but one of many examples. In this chapter we discuss several problems including Brownian motion and thermal distributions.

8.1   Random Numbers and Nuclear Decay

The first thing we need to simulate random processes is a random number generator. Because digital computing relies on precise digital logic, we can generate only pseudo random numbers that mimic some properties of true random numbers (unlike quantum computing, see Chap. 7).

The details of pseudo random number generation need not concern us, suffice it to say that it involves multiplication of large integers to cause an overflow, and the overflown bits look rather random, and they are used to form pseudo random numbers.

A basic random number generator outputs a number between 0 and 1 uniformly. Python has a random number module, but Numpy has a more versatile and convenient random number library. Specifically, `random()` produces a single random number, and `random(N)` an array of N random numbers. Sometimes we want integer random numbers. The function `randint(m, n, N)` will output an array of N integer random numbers between m (inclusive) and n (exclusive). If N is omitted, it will output just a single number between m and n.

© The Author(s), under exclusive license to Springer Nature Switzerland AG 2023   175
J. Wang and A. Wang, *Introduction to Computation in Physical Sciences*, Synthesis Lectures
on Computation and Analytics,
https://doi.org/10.1007/978-3-031-17646-3_8

Nuclear Decay

Let us simulate the nuclear decay problem, namely, the decay of unstable particles. The problem will help us get familiarized with the use of random numbers.

Nuclear decay is fundamentally governed by quantum physics (Chap. 7), and is a true random process. Let $N(t)$ denote the number of nuclei remaining at time t, and $N(t + \Delta t)$ later at $t + \Delta t$. In the time interval Δt, some nuclei will decay, so the change of remaining nuclei is $dN = N(t + \Delta t) - N(t)$. Because the number of nuclei will decrease, the value dN will be negative, i.e., $dN < 0$.

What does dN depend on? We expect it to be proportional to two factors: $N(t)$ and Δt, because more existing nuclei means more will decay, and a larger time interval also means a greater chance of decay. So we can write the following relationship, assuming λ as the proportional constant,

$$dN = -\lambda N \Delta t. \tag{8.1}$$

The negative sign in Eq. (8.1) ensures that the change of nuclei is negative. Looking at the relationship another way, we can rewrite it as

$$\frac{-dN}{N} = p, \quad p = \lambda \Delta t.$$

Here p is the fraction of nuclei that has decayed in the interval, so we can interpret p as the probability with which a nucleus will decay in the time interval Δt.

With this interpretation, we can construct simulation of nuclear decay this way: During an interval, for each of the remaining nuclei, we compare the probability p with a random number x. If $x < p$, the nucleus decays, otherwise it survives. Because x is uniform between 0 and 1, statistically the fraction of decayed nuclei will be equal to p. The following code illustrates core idea, taken from Program A.6.1.

```
In [1]:  dN = 0
         for _ in range(N):
             if random() < p:
                 dN = dN - 1
         N = N + dN
```

Figure 8.1 shows the results of a nuclear decay simulation from Program A.6.1. The linear scale shows that N decreases with time. Though exponential decay is expected from Eq. (8.1), it is not at all clear on the linear scale. To verify it, we plot the data on the semilog scale (Fig. 8.1, right). It shows a straight line for the most part, confirming it is indeed exponential.

There is scatter in the data, due to the statistical fluctuations. The scatter is bigger near the end on the semilog scale, as expected, because smaller N means larger relative fluctuations. The scatter near the end is barely noticeable on the linear scale where large numbers dominate. It is clearly visible only on a log scale where both the small and the large are

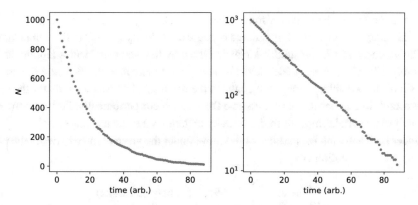

Fig. 8.1 Decay of unstable particles as a function of time on a linear scale (left) and semilog scale (right). The initial N is 1000, and the probability $p = 0.05$

represented on equal footing. The takeaway is that when presenting numbers across orders of magnitude, we should consider the log scale. Of course, this is not the only reason. It is also the preferred scale to reveal exponential trends or power scaling laws (see Sects. 8.2 and 8.2.2).

8.2   Brownian Motion

You may have seen how a drop of food coloring spreads in water or how a bit of fine powder feed quickly disperses in a fish tank. Actually the naked eye misses a lot of important details. If we could observe the process filmed through a microscope and in slow motion, we would see the particles move about randomly, and frequently execute jerky, sudden acceleration as if being kicked by an invisible force. The motion of small macroscopic particles in liquids is called Brownian motion, which is often observed in biophysical systems. The importance of Brownian motion exceeds far beyond physics or biology, with applications in mathematics, finance and related stochastic models. Here we aim to understand Brownian motion with simulations sans expensive equipment.

8.2.1   A Random Kick Model

Let us develop a model of Brownian motion, starting with forces acting a particle. A particle immersed in water interacts with a great many number of water molecules (assuming gravity is balanced by buoyancy), so treating each and every particle-molecule interaction is practically impossible. Rather, we should consider collective or statistical effects.

The first effect is the viscous force, $F_d = -b_1\mathbf{v}$, akin to the linear drag Eq. (5.12). The quadratic drag force can be safely ignored because the velocity is small in Brownian motion.

The second effect is the sudden kicks delivered by the water molecules zipping around randomly. Modeling these sudden kicks is nontrivial. In any model, we would like to keep it as simple as possible, as long as it contains the essential physical elements for the model to be useful. In that spirit, we will assume the kicks occur periodically. Furthermore, each kick delivers a certain impulse so the velocity undergoes a sudden change $\Delta\mathbf{v}$.

Under such a model, the particle would move under the viscous force between the kicks. The equations of motion are,

$$\frac{d\mathbf{x}}{dt} = \mathbf{v}, \quad \frac{d\mathbf{v}}{dt} = -b\mathbf{v}, \quad \text{(in-between kicks)} \tag{8.2}$$

where $b = b_1/m$. In terms of the initial position and velocity $\mathbf{r}_0$ and $\mathbf{v}_0$, respectively, the solutions are

$$\mathbf{v} = \mathbf{v}_0 \exp(-bt), \quad \mathbf{r} = \mathbf{r}_0 + \frac{\mathbf{v}_0}{b}[1 - \exp(-bt)]. \tag{8.3}$$

If Δt is the interval between kicks, the velocity will decrease exponentially during the interval to $\mathbf{v}_0 \exp(-b\Delta t)$, at which point it will be changed by $\Delta\mathbf{v}$ after a kick at the beginning of the next interval. We can thus treat the motion as a discrete map as follows. Letting $\mathbf{r}_n$ and $\mathbf{v}_n$ denote the respective position and velocity at the beginning of the current interval, their values at the beginning of the next interval (right after the kick) are

$$\mathbf{r}_{n+1} = \mathbf{r}_n + \frac{\mathbf{v}_n}{b}[1 - \exp(-b\Delta t)] \quad \mathbf{v}_{n+1} = \mathbf{v}_n \exp(-b\Delta t) + \Delta\mathbf{v}. \tag{8.4}$$

We can march forward with Eq. (8.4) one interval at a time, with each pair of $(\mathbf{r}_{n+1}, \mathbf{v}_{n+1})$ serving as the initial conditions of the next interval, and so on.

8.2.2   Spatial and Temporal Evolution

Now that we have developed a model of Brownian motion, how does it work, considering its simplicity? The simulation Program A.6.2 implements the method Eq. (8.4) in 2D and the results are shown in Fig. 8.2. All trial particles are released at the origin from rest. They spread outward and form clusters of different sizes at different times.

To keep it simple, the impulse $\Delta\mathbf{v}$ is assumed to be constant in magnitude but random in direction. This is done by randomly choosing an angle φ between $[0, 2\pi]$, and calculating the impulse as $\Delta\mathbf{v} = |\Delta\mathbf{v}|(\cos\varphi\,\hat{i} + \sin\varphi\,\hat{j})$. The simulation runs forward one step, applies a random impulse, and repeats. A procedure based on such random samplings is called a Monte Carlo method or simulation. A Monte Carlo method approaches a problem by breaking it down to simple, basic rules; and then applying the rules repetitively with proper sampling to statistically build up solutions of likely outcomes. This method is a valuable tool in scientific computing, especially in solving complex problems that are otherwise

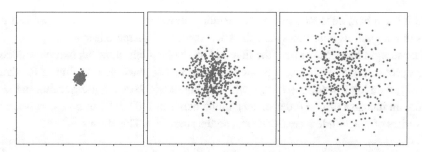

Fig. 8.2 Cluster of Brownian particles at different times: near the beginning (left, 50 steps), intermediate time (middle, 250 steps), and large time (right, 500 steps). The number of trial particles is 500

intractable. Its approach is also a reflection of how complex systems are built from simple rules (Chap. 9).

Figure 8.2 shows that the particles spread out in all directions nearly uniformly. Aside from fluctuations, the average position of the particles is expected to be zero because of random motion. However, the rate of spread (growth of the cluster size) is much faster at the beginning, and slows down considerably at larger times. This can be seen more clearly from watching the animation of Program A.6.2.

The rate of the spread is related to the growth rate of the cluster size. How do we characterize the size? Because the average position is zero, we need to consider the next moment, the average of position squared, namely, the average distance squared as a measure of the cluster size. We can also obtain this information from Program A.6.2. We leave the calculation as an exercise (see Exercise 8.1) and show the results in Fig. 8.3.

The average position squared, $\langle \mathbf{r}^2 \rangle$, is shown on a log-log scale (Fig. 8.3, left). There are roughly three regions. The first region starts from zero and increases rapidly at the beginning.

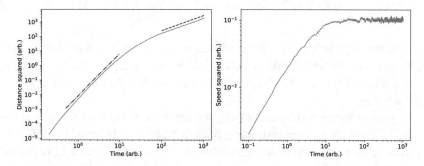

Fig. 8.3 Average distance squared (left) and average speed squared (right) as a function of time. The number of trial particles is 500. The dash-dot and dashed lines are drawn to show the approximate slopes in the respective regions

Then it has a bent knee structure at intermediate times around $t \sim 5 \times 10^1$, we call it the transition region. At later times after the knee, the rate of increase is slower.

The straight lines drawn over the first and the last regions show approximate slopes of the respective regions because a log-log plot can reveal power-law scaling in the data (if $y = cx^k$, then $\log y = \log c + k \log x$ with k being the slope on a log-log scale). In the first region, an increase of one order of magnitude between $[10^0, 10^1]$ in time corresponds to two orders of magnitude increase in $\langle \mathbf{r}^2 \rangle$, so the slope is 2. This means

$$\langle \mathbf{r}^2 \rangle \propto t^2, \quad \text{or } r_{rms} = \sqrt{\langle \mathbf{r}^2 \rangle} \propto t. \quad \text{(pre-transition)} \qquad (8.5)$$

This says that the cluster size r_{rms} grows linearly with time initially.

After the intermediate transition region, the rate of increase shows again a power law. Now an increase of one order of magnitude between $[10^2, 10^3]$ in time results in one order of magnitude increase in $\langle \mathbf{r}^2 \rangle$, a proportional relationship like

$$\langle \mathbf{r}^2 \rangle \propto t, \quad \text{or } r_{rms} = \sqrt{\langle \mathbf{r}^2 \rangle} \propto \sqrt{t}. \quad \text{(post-transition)} \qquad (8.6)$$

Comparing Eq. (8.5) with Eq. (8.6), we find the initial rate is much greater than the final rate. To double the r_{rms} size, the time interval needs to also double before the transition, but it must be 4 times as long after the transition. The latter is a universal large-time behavior in stochastic expansion, including diffusion.

We now turn our discussion to the average velocity squared $\langle \mathbf{v}^2 \rangle$ (Fig. 8.3, right). It has three regions as well, an initial rapidly climbing region, a relatively sharp transition, and a flat post-transition. Like $\langle \mathbf{r}^2 \rangle$, the pre-transition region follows a power law, but the power is close to 1, so $\langle \mathbf{v}^2 \rangle \propto t$. We may understand this scaling law in terms of impulse received. Initially the velocity is very small, so we can neglect the viscous force and focus only on the the kicks. After n kicks, the velocity is the sum of all kicks $\mathbf{v} = \sum_i \Delta \mathbf{v}_i$. The velocity squared is given by

$$v^2 = \mathbf{v} \cdot \mathbf{v} = \sum_{i,j} \Delta \mathbf{v}_i \cdot \Delta \mathbf{v}_j = \sum_i \Delta \mathbf{v}_i^2 + \sum_{i \neq j} \Delta \mathbf{v}_i \cdot \Delta \mathbf{v}_j. \qquad (8.7)$$

Because the magnitude of each impulse is assumed to be the same in our model, $|\Delta \mathbf{v}_i| = |\Delta \mathbf{v}|$, the sum of square terms add to $|\Delta \mathbf{v}|^2 \times n$. The sum of cross terms would add up to zero on average because the impulses are random in direction. So we have $\langle \mathbf{v}^2 \rangle \propto n \propto t$ in the initial region.

In the post-transition region, the velocity has increased to the point where the viscous force Eq. (8.2) cannot be neglected. The gain from the kicks is balanced out with the loss from viscosity on average, so the average velocity squared remains flat, apart from statistical fluctuations (Exercise 8.1).

Overall, despite its limited simplicity, our model works surprisingly well in reproducing the physical properties of Brownian motion and, more important, helping us gain insight and deeper understanding of the phenomenon.

8.3 Thermal Equilibrium

In the preceding section we treated the Brownian particles apart from the atoms in liquid (bulk). But in most thermal systems, it is the interactions of atoms or molecules among themselves that determine the thermal properties of the bulk material. In this section we discuss simulations of thermal energy exchange, thermal equilibrium, and equilibrium distributions within a simplified model. We will look at how these microscopic interactions affect macroscopic properties.

8.3.1 A Thermal Interaction Model

It is helpful to think of a generic thermal system as being made up of atoms (or molecules) of the same kind. We can visualize these atoms interacting with each other, exchanging energy when an interaction occurs (e.g., collisions or vibrations). To keep it simple, we can imagine the atoms as little balls rolling around, carrying integer amount of energy. They either gain or lose one unit of energy (discrete energy levels, a consequence of quantum mechanics, see Chap. 7) when they touch each other.

As a thermal model, consider a thermally isolated system consisting of N atoms, each carrying discrete amount of energy but the total energy is constant. The lowest amount of energy an atom can have is zero, and there is no upper limit. Any pair of atoms in the system can interact with each other. When an interaction occurs, one atom can gain one unit of energy from, or lose one unit to, the other atom. The probability of gaining or losing energy is equal. This is called the principle of statistical postulate. It states that all internal configurations of a system consistent with its external properties are equally accessible. In this model the configurations are the possible ways of distributing energy among the atoms, and the external property is the total energy. The model thus described is an entirely egalitarian, equal energy-sharing model. It is also called an Einstein solid.

If all atoms start out with the same amount of energy, and if the rules of interaction are equal and favor no particular atom, what results would we expect? Would every atom end up with more or less equal amount of energy after some interactions, or something else? (Also see Exercise 8.2.) It would be a good time for us to pause here, think about the question, and make a prediction (and write it down). Let us find out the answer next.

8.3.2 Thermal Energy Distribution

We now consider a Monte Carlo simulation of an Einstein solid. We assume the solid has N atoms and is thermally isolated. Interactions occur only among the atoms in the system, and its total energy is conserved. We can represent the atoms with a list (array) of size N. The (integer) value of the n-th element in the list gives the energy of atom n. We can initialize the list, then randomly iterate through pairs of elements (i, j), exchange one unit of energy by adding 1 to element i and subtracting 1 from element j. This model is implemented in Program A.6.3.

It may seem unrealistic that we could compute the thermal properties of any system by following just N particles, because whatever N is allowable on a digital computer, it will be so minuscule as to be utterly negligible relative to actual thermal systems with vastly many more particles. In fact and for this very reason, we specifically avoided treating each and every particle-molecule interaction in Brownian motion (Sect. 8.2) regarding the effects of the bulk background (liquid). Rather, we considered only statistical effect of the liquid, viscosity. As we shall see, this approach can and does work, if we are careful with doing the simulation in a statistically meaningful way.

Now, we are ready to look at the energy distribution in terms of the number of interactions, namely, time. Figure 8.4 displays the results from the simulation Program A.6.3. The atoms ($N = 256$) are arranged in a square lattice for easy visualization, with each lattice site representing an atom and its color signifying the amount of energy it has.

All the atoms were initialized to the same amount of energy. In the early stage (Fig. 8.4, left), not many interactions have occurred yet, and the variation in energy is visible but not very significant. Most sites are in low-energy, blueish color. The situation changes a lot at the intermediate stage where we see large variation in the energy, a few high-energy bright sites start to appear, along with medium-energy ones. Still there are many low-energy darker sites.

At the last stage (Fig. 8.4, right) after many more interactions, there are more high-energy sites than earlier, but their number is dwarfed by medium- to low-energy sites. In fact the latter still dominate the lattice. The picture is that there are a few atoms having a lot of energy (6 or 7 out of 256), but the vast majority of atoms have less energies now than at the beginning. So a haves-and-havenots class divide is created, based on equal-energy sharing rules. (What was your prediction?)

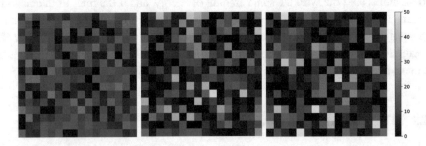

Fig. 8.4 Energy distribution from a Monte Carlo simulation of a solid ($N = 16 \times 16$) at different times: near the beginning (left), intermediate time (middle), and large time (right). Initially all atoms have the same, 10 units of energy

Continuing with more interactions and energy exchanges does not change the last-stage distribution much. The high-energy atoms may be different (locations of bright sites would change), and low-energy atoms could acquire more energy, but their relative numbers do not change significantly. We say the system has reached equilibrium.

Boltzmann Distribution

Having discussed the energy distribution qualitatively, we now wish to characterize it quantitatively. Specifically, what exactly are the ratios, or probabilities, of atoms with certain amount of energy to the total population N?

Let c_n be the count of atoms with energy E_n, then $P_n \equiv P(E_n) = c_n/N$ is equal to the probability of finding an atom with energy E_n. A slight modification to the simulation Program A.6.3 can be made to tabulate the number of atoms with certain energy (see Exercise 8.3 or Program A.6.4). After many interactions and the system is presumably in equilibrium, the probabilities are calculated and shown in Fig. 8.5.

The probabilities vary over 3 orders of magnitude in the energy range [0, 10], so the results are plotted on a semilog scale for similar reasons discussed following Fig. 8.1. Aside from the scatter toward the end, most data points fall neatly in a straight line, indicating an exponential distribution,

$$P(E_n) = \exp(-\beta E_n). \quad \text{(Boltzmann distribution)} \tag{8.8}$$

Equation (8.8) is the well-known Boltzmann distribution, and is the foundation of statistical mechanics. It can be derived vigorously using the principle of statistical postulate, with $\beta = 1/kT$ where k is the Boltzmann constant and T the absolute temperature. It is remarkable that our tiny model is able to so accurately demonstrate the Boltzmann distribution. This is in part because the atoms are sampled in a statistically meaningful way, and also because the Boltzmann distribution is purely a probabilistic statement based on the statistical postulate regardless of the nature of interactions.

Fig. 8.5 The probability as a function of energy of an atom in a solid in equilibrium, corresponding to the large time limit of Fig. 8.4. It is normalized to be one at $n = 0$. The straight line represents an exponential distribution $\exp(-\beta n)$, with $\beta = \ln 2$

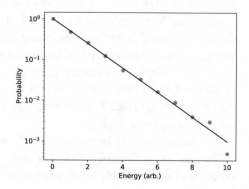

The slope in Fig. 8.5 is determined by the parameter β. A straight line is drawn in Fig. 8.5 using Eq. (8.8), and we find that $\beta = \ln 2$ best fits the data. In our solid model, kT depends on the average energy of the atoms, which is set to be one in the simulation. It is shown elsewhere that this lead to a $kT = 1/\ln 2$ energy units, hence $\beta = 1/kT = \ln 2$ gives the best fit (Exercise 8.3).

A Force Toward Equilibrium: Entropy

The simulation and results above show that the solid evolves from an initially uniform energy distribution to a very uneven equilibrium distribution characterized by the Boltzmann distribution. By equilibrium, we mean the solid has reached a state where the Boltzmann distribution remains unchanged within statistical fluctuation.

We can ask a question: is the reverse possible? That is, will the system spontaneously go back to its initial uniform configuration at some point? According to the principle of statistical postulate stated earlier, every accessible state is equally probable, and the uniform configuration is an accessible state. So in that sense, it is possible. But is it probable? Not really. Given the many possible configurations even for our very tiny solid, the probability will be so small as to be practically impossible. There is not enough time in the universe for the reverse to happen. We will come back to this point shortly.

Now, if the thermal process is irreversible (without external interference), what drives the system toward equilibrium? How can we be sure that equilibrium has been established? The clue is found in terms of entropy, a physically measurable quantity in statistical mechanics (see Exercise 8.6). Entropy can be calculated in different but equivalent ways, one of which is given by the following expression (see Blundell and Blundell (2010)),

$$S = -k \sum P_n \ln P_n, \quad \text{(entropy)} \tag{8.9}$$

where P_n is the probability of finding the system in state n (another definition is given in Exercise 8.4).

For our solid model, the probability can be taken from the Boltzmann distribution Eq. (8.8), and Eq. (8.9) will be convenient for calculating entropy from the simulation. Incidentally, the scaled entropy $s = S/k$ is defined as information entropy outside of statistical mechanics. A uniform distribution like the initial state would mean zero entropy because one state has unitary probability and all others have zero probabilities.

Figure 8.6 displays the entropy of an atom obtained from Program A.6.4 as a function of the iteration number, namely time. It is plotted on a semilog horizontal scale because the entropy changes rapidly in the beginning but very slowly at the end. The semilog scale accentuates both ends equally.

The entropy starts from zero because $P_0 = 1$ and $P_{n \neq 0} = 0$, so the initial state conveys no information as discussed earlier. It rises very quickly as soon as interactions start to occur. Afterward, it slowly levels off near the end and reaches a plateau value and does not

Fig. 8.6 Scaled entropy S/k of an atom as a function of iteration number (time). The solid system is the same as in Fig. 8.4

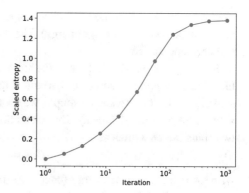

change appreciably thereafter, apart from possible fluctuations. This signals that the system has reached equilibrium when entropy is maximum.

These observations from the simulation lead us to the conclusion that entropy tends to increase, and it is at the maximum in equilibrium. This statement is in fact the second law of statistical mechanics. In this sense, we can view entropy as an arrow of time (Exercise 8.4).

Now we can revisit the question posed earlier about reversibility of the system. There we made an argument that it is mathematically possible but improbable for the system to go back to the initial uniform configuration. In terms of physics, we can answer the question much stronger: it is not possible for the reserve process to occur spontaneously. It would imply that entropy would decrease, and that would violate the second law of thermodynamics.

8.4   The Ising Model

In both Brownian motion (Sect. 8.2) and the solid model just discussed, the atoms are independent of each other. They each gain or lose energy regardless of what states other atoms are in. There are systems where the particles are interdependent (or coupled) and the state of one particle affects other particles. For example, in a magnet, magnetic dipoles of iron atoms in close neighborhoods align themselves spontaneously in the same direction, producing a nonzero magnetization, called ferromagnetism, even absent an external magnetic field. We discuss the simulation of such a coupled system known as the Ising model, and explore its thermal behavior and properties.

8.4.1   The Spin Chain

When atoms are placed next to one another like in a lattice, electronic clouds (quantum wave functions, see Chap. 7) can overlap with each other. Usually the overlap causes an energy difference depending on the orientations of the atoms, resulting in a higher or lower electronic energy if they are parallel or antiparallel, or vice versa (resulting net repulsion or attraction, known as exchange energy). The details of this interaction can be quite complex and need not concern us.

For our purpose, we can think of the atoms as spins (a fundamental property of particles like electrons, protons, and so on) arranged in 1D a chain (see Fig. 8.7). Let s_i denote the spin of the ith atom. To keep it simple, we assume each spin can point either up or down, so $s_i = 1$ (up) or $s_i = -1$ (down). The orientation-dependent interaction energy between any two spins can be written as

$$E_{i,j} = -\epsilon\, s_i s_j, \tag{8.10}$$

where ϵ is the pair-wise energy (coupling). If $\epsilon > 0$, the interaction favors parallel spins (ferromagnetism). In our discussion, we assume $\epsilon > 0$, treating it as the basic unit of energy. It is understood that energy and temperature kT are expressed in terms of this unit. A negative $\epsilon < 0$ is also possible, in which case antiparallel spins are preferred (antiferromagnetism, Exercise 8.7).

Though all pairs can interact with each other, the Ising model assumes only nearest neighbor interactions, neglecting weak, long-range interactions. We can express the total energy of the 1D Ising model as the sum of adjacent pairs

$$E = \sum_i E_{i,i+1} = -\sum_{i=1}^{N} s_i s_{i+1}. \quad \text{(in units of } \epsilon) \tag{8.11}$$

We run into a problem with the last term because s_{N+1} does not exist. Instead of just neglecting this term, a better way is to use the so-called periodic boundary conditions,

$$s_{N+1} = s_1. \quad \text{(spin ring)} \tag{8.12}$$

We can visualize Eq. (8.12) as a spin chain bent into a ring, so there is no beginning or end and no unpaired spins. Periodic boundary conditions can minimize finite size effects in this case, but they are also useful in other simulations like waves (see Exercise 6.7).

The total energy E can vary from $-N$ (perfect alignment) to zero (random alignment). If all spins were reversed simultaneously $s_i = -s_i$, there will be no change in the total energy. This is because there is no preferred direction in the absence of external fields.

8.4.2 Monte Carlo Simulation

Internal interactions will not change macroscopic properties of a spin chain such as energy or magnetization. To effect such change, it must be in thermal contact with an external environment. The environment is assumed to be large (often called a heat bath) so its temperature remains constant while exchanging energy with the Ising system.

Let us picture a spin chain in thermal contact with a heat bath at temperature T. Energy can be exchanged via collisions or vibrations but, as seen in the solid model, the specific mechanism is unimportant. Just like atoms in the solid model (Sect. 8.3.1), the spins will flip to reflect energy absorbed or given off, driving the Ising system toward equilibrium.

Intuitively we expect that if the energy of the system Eq. (8.11) is too low, it has the tendency to absorb energy from the heat bath, causing the spins to align antiparallel to each other to store the increased energy. However, if the energy is too high, it will tend to give off energy to the heat bath, causing the spins to flip toward parallel alignment. These spin flips will continue until the system reaches equilibrium after sufficient time passes. After that, random flips can still occur, but on average, parallel and antiparallel flips balance each other out. Eventually the system is expected to follow the Boltzmann distribution Eq. (8.8), with energy, and magnetization, fluctuating around equilibrium values.

The question facing us is: how do we realize the process in a simulation? Recall in the solid model (Sect. 8.3.1), atoms randomly exchange energy with each other, and the Boltzmann distribution comes naturally as a result. We could attempt the same thing here, but because the Ising chain is exchanging energy with the heat bath instead of with another Ising chain, the process cannot be totally random. If it were random, the system would gain or lose energy with equal probability because the heat bath is impervious to energy absorbed or given off. The Ising chain would never reach equilibrium that we intuitively expect. A proper bias needs to be applied.

As it turns out, a method known as the Metropolis algorithm (see Metropolis et al. (1953)) is precisely what is needed here. It applies a proper bias to efficiently nudge the system toward equilibrium. Using the Ising model as an example, the bias is applied in the following manner (see Gould and Tobochnik (2021)). Starting from a given configuration, a spin is randomly chosen for a trial flip. Let ΔE be the energy difference caused by the flip. If the flip yields a lower energy, i.e., $\Delta E < 0$, the trial flip is accepted. If, however, the trial flip leads to a higher energy, $\Delta E > 0$, it is not rejected outright. Rather, it is accepted with a probability $\exp(-\Delta E/kT)$ according to Eq. (8.8).

The reason for accepting trials with $\Delta E < 0$ fully is because lower energy states dominate the Boltzmann distribution (Fig. 8.5). On the other hand, higher energies above the average value is still thermally possible, albeit with much reduced probabilities, so accepting some trials with $\Delta E > 0$ is consistent with the Boltzmann distribution. It will also ensure that thermal fluctuations can occur.

The Metropolis method is utilized in Program A.6.5 and select configurations are shown in Fig. 8.7 at several points in temperature. At the heart of the simulation is the way the energy difference ΔE is efficiently calculated due to a trial flip. Because a single flip affects energies with the nearest neighbors only, it is possible to find ΔE efficiently without summing up all pairs from Eq. (8.11).

Suppose spin s_n is to be trial-flipped. Before the flip, the residual energy from its left and right neighbors is $-s_{n-1}s_n - s_n s_{n+1}$. After the trial flip $s_n \to -s_n$, the residual is just negative of that, so the net change of energy is twice of the latter,

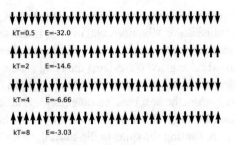

Fig. 8.7 The Ising chain ($N = 32$) at select points from low to high temperature. The total energy is shown for each configuration. The energy would be the same if all spins were reversed at once because there is no preferred direction. It is understood both the energy E and temperature kT are in the basic (arbitrary) unit ϵ characteristic of the pair-wise interaction energy

$$\Delta E = 2s_n(s_{n-1} + s_{n+1}). \tag{8.13}$$

There is no energy change if the nearest neighbors are antiparallel. The most negative change occurs ($\Delta E = -4$) when the trial flip aligns the spin parallel to the nearest neighbors. The opposite causes the most positive change ($\Delta E = 4$).

With Eq. (8.13), the Metropolis method may be summarized as follows: Randomly pick a spin n and compute ΔE; If $\Delta E \leq 0$, accept the flip, else accept it with a probability $\exp(-\Delta E/kT)$. Once accepted, update the spin state. Below is part of the code taken from Program A.6.5.

```
In [2]:  accept = False
         if (dE <= 0.0): accept = True    # flip if dE<0
         else:
             p = np.exp(-dE/kT)                # else flip with exp(-dE/kT)
             if (random() < p): accept = True
         if accept:                            # actual flip
             E = E + dE
             spin[n] = -spin[n]
```

It is convenient to represent and manipulate the Ising chain as a list (array) spin. The possible values of each element are ± 1. The variables kT and E are the temperature and energy, respectively, of the current state. The first if-block decides whether the trial flip is to be accepted according to the Metropolis algorithm, and the second if-block actuates the flip if accepted.

Note that instead of using T itself, it is more convenient to use kT in Program A.6.5 because it has the dimensions of energy, which can be expressed in the energy unit ϵ (Eq. (8.10)), so it is a pure number, just like E. Also, another reason is that it is always possible to pair energy and temperature as a ratio of E/kT (or ϵ/kT) in thermal simulations. Energy and temperature scales are inherently interconnected in a given physical system.

The results in Fig. 8.7 are snapshots after equilibrium at each temperature, which is controlled by a slider in the simulation. It is important to sample the system after many iterations following a temperature change to ensure equilibrium is reached. Even then, the average values such as the energy should be calculated over a sufficient number of samplings, given the system size (1000 used in this case) and thermal fluctuations. The snapshots show transitions from perfect spin alignment at low temperature ($kT = 0.5$) to totally random alignment (equal up and down spins) at high temperature ($kT = 8$). The two configurations at intermediate temperature ($kT = 2$ and 4) have a similar up/down spin ratio but different domain sizes, resulting in more antiparallel spins at the domain boundaries and a large difference in the energy as temperature is increased.

Running the simulation, we can observe that once the perfect alignment is attained at low temperature, it is very stable. The probability is very low for a spin to flip out of this configuration because the energy cost is high according to Eq. (8.13). For instance, if a spin is to flip at $kT = 0.5$, the energy difference would be $\Delta E = 4$, and the probability of acceptance is $\exp(-\Delta E/kT) \sim 3 \times 10^{-4}$, very rare.

It would also seem that full magnetization (average spin being 1) is possible at low temperature because of perfect alignment. But this is misleading. If we raise the temperature, the perfect alignment will be broken. If we then lower the temperature again, the perfect alignment will happen, but it is not certain that the direction would be downward as shown in Fig. 8.7. Statistically, it has an equal chance of pointing up or down, so the net magnetization of the 1D Ising model is in fact zero.[1] It turns out this is true at any temperature due to lack of an external field (see Exercise 8.5).

Figure 8.8 shows the average energy over a larger range of temperature. The results are obtained from Program A.6.5 modified to compute the equilibrium average over sufficient number of samplings as noted earlier (see Exercise 8.5). Also, to reduce fluctuation, a larger Ising chain ($N = 400$) is used.

In contrast to zero magnetization discussed earlier, the average energy shows different behaviors in three regimes of temperature. Because the regimes cover a large range of temperature, Fig. 8.8 is plotted on the semilog temperature scale. At the low-temperature regime, it is flat and approaches -1 per spin, due to all spins in perfect alignment seen earlier (Fig. 8.7). There is a transition regime at intermediate temperature where the average energy rapidly rises with increasing temperature, corresponding to the formation of smaller domains with antiparallel spins (Fig. 8.7). In the last regime at large temperature, the energy increases very slowly toward zero.

We note a few related points for further exploration. First, an important property we have not discussed is the heat capacity, defined as the ratio of change of energy to change of temperature. Simply stated, it is the slope of energy as a function of temperature, a

---

[1] The simplicity of the 1D Ising model makes it an ideal system for studying nontrivial thermal properties. But its limitation is that the model does not exhibit phase transition, an abrupt phenomenon where full magnetization occurs below a critical temperature. However, the Ising model in 2D does show phase transition (see Schroeder (2021)).

Fig. 8.8 The average energy per spin of an Ising chain ($N = 400$) as a function of temperature

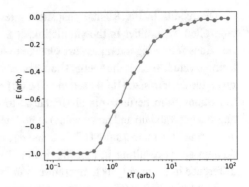

measure of the energy needed to raise a system's temperature by certain amount. We can see from Fig. 8.8 that the heat capacity will be zero at both low and high temperature, and peaks somewhere in the intermediate regime. We leave its exploration to Exercise 8.5. The second point of interest is the entropy of the Ising model. Instead of calculating it from a formula, we discuss a way of "measuring" entropy from the heat energy entering the system in Exercise 8.6. Lastly, we mentioned antiferromagnetism earlier due to antiparallel alignment of spins when the pair-wise coupling is negative $\epsilon < 0$. In this case, the so-called staggered magnetization is a useful measure of order. Other interesting properties may be further investigated in Exercise 8.7.

8.5   Energy Balance and Global Warming

We have discussed several thermal models exchanging energy with a large environment called heat baths. The largest of them all would be mother Earth for all living creatures including humanity. Earth is heated by the sun during the day, and excess heat is radiated back into space mostly during the night. A delicate balance is required to maintain a hospitable environment.

Since the industrial revolution, however, the balance has been and is still being upset. The amount of heat trapping gases, primarily CO_2, has increased dramatically in the atmosphere, blocking radiation of excess heat back into space. This causes a rising temperature on Earth's surface, namely global warming, leading to melting of ice caps, extreme weather, and rising sea levels that pose existential threat to humanity. As of this writing, it was reported in late March, 2022 that temperatures reached unheard extremes of 40 °C and 30 °C above normal at Antarctica and the north pole, respectively, and simultaneously.

Scientific evidence points to man-made global warming at an alarming rate relative to natural cycles that occur over tens of thousand of years. This is evident from the temperature change data shown in Fig. 8.9. The baseline for comparison is 1850. The pre-1850 data is obtained from a combination of scientific modeling and reconstruction. The post-1850

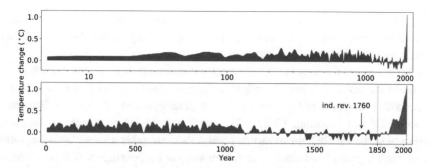

Fig. 8.9 IPCC data on change of Earth's surface temperature as a function of time relative to temperature at 1850. In the upper figure, the time axis is on the log scale. The dotted and dashed vertical lines in the lower figure indicate respective years 1850 and 2016, the last year the data is available. The red colored curve between 1850 and 2016 is for emphasis. The current sharp rise actually started around 1920, merely a century ago, 70 years after the reference point of 1850 (dotted line). The arrow marks 1760, the start of the industrial revolution. The current rise in temperature is 1.1 °C relative to the preindustrial era. Source: IPCC Dataset, Ref. Gillett et al. (2021)

data comes from observational data and records. Over the last 2000 years, the temperature fluctuates around zero until about 1920, when the current, sustained sharp rise in temperature started and is continuing. Both the logarithmic and linear time scales in Fig. 8.9 accentuate the sudden uptick. Currently, the rise in temperature is 1.1 °C relative to the preindustrial era. Lest anyone thinks this is a small warming in the abstract, it represents an enormous increase of energy on Earth's surface (see next), capable of influencing or even altering well established global circulation patterns, causing more frequent extreme weather events and wreaking catastrophic climate change unless globally coordinated action takes place immediately. The dramatic rise over the last century is accompanied with an equally dramatic rise in carbon dioxide (CO_2) concentration in the atmosphere (not shown here; see NASA (2022) for the latest measurement), and cannot be explained by scientifically known, naturally occurring phenomenon.

The short amount of time over which the rise occurs is a mere blip in the history of humanity, much less history of Earth. In fact, a recent study shows that (see Miller (2022)) "never in the past 24 millennia has Earth been warmer than it is today, and never has it warmed faster than it is warming today." More heat is being trapped because of emission of greenhouse gases due to man-made activities, greatly accelerating the rate of warming.

Energy Balance

The physics of energy balance on Earth's surface is broadly well understood, albeit with some precise details still being investigated scientifically. The incoming sunlight energy on Earth is transferred in many different forms including: direct reflection back into space, heating of the surface (mostly oceans), vaporization of water, absorption via excitation or ionization of atmospheric molecules, photosynthesis, thermal radiation, greenhouse trapping, and so on (see Trenberth et al. (2009)). All these mechanisms effectively reach a thermal equilibrium, maintaining Earth's surface temperature at T.

In thermal physics, a body at temperature T will radiate thermal energy, known as black-body radiation, which consists of continuous electromagnetic waves peaked at a frequency proportional to temperature T given by Wien's displacement law (see Sect. 3.3, Exercise 3.4, also see Gould and Tobochnik (2021)). The sun is such a blackbody source radiating thermal energy in the visible light spectrum, though there are other sources including infrared emitters (holding your hand close to the face or a cup of hot beverage will attest to this fact).

The blackbody radiation of the Earth would have escaped into space unimpeded if not for the presence of greenhouse gases such as CO_2. The fact that CO_2 is a heat-trapping greenhouse gas is well understood physically. Typically, gas molecules can store energy via translational motion (kinetic energy) and vibrational motion.[2] The latter is important for understanding the greenhouse effect. We have already seen the classical oscillations of triatomic molecules like CO_2 (see Sect. 6.1.3 and Eq. (6.13e)). At the molecular level, molecules absorb energy via excitation of quantum vibrational states like the simple harmonic oscillator (Sect. 7.3.3). Owing to its molecular structure, the vibrational excitation energies of CO_2 correspond to wavelengths in the range 2–15 microns (10^{-6} m) in the infrared region of optical spectrum. This is the key to understanding greenhouse heat trapping and its role in energy balance.

The Earth at the nominal room temperature ($T \sim 300$ K) emits blackbody radiation with peak wavelength at $\sim$10 microns, closely matching the vibrational excitation energies of CO_2 (see Exercise 8.8). As the blackbody radiation flux propagates outward through the atmosphere, some portion of it will escape into space directly. The rest will encounter air molecules such as oxygen (O_2), nitrogen (N_2), carbon dioxide (CO_2), and so on. However, the vibrational states of O_2 and N_2 are outside the infrared region for excitation, so they appear transparent to blackbody radiation from Earth. This is because quantum transitions occur only if the photon energies match the excitation energies, and otherwise no interactions at all.

The situation is different with CO_2 and similar greenhouse gases that can absorb infrared photons (see Fecht (2021)). After absorbing the photons and being excited to higher vibrational states, the CO_2 molecules will decay back to the lower states by emitting photons of the same energy. Now, these re-emitted photons will be distributed randomly in all directions, approximately half back to Earth and half to outer space. The re-emission of radiation back to Earth is the heat-trapping, greenhouse effect. Without the greenhouse effect, Earth would be too cold, so the presence of certain amount of greenhouse gases like CO_2 is good and necessary for Earth to maintain its hospitable temperature.

However, when there is increasingly more concentration of greenhouse gases in the atmosphere, the amount of heat trapped increases, causing the surface temperature to rise.

---

[2] Rotational motion can also store energy, but it is open only at much higher temperatures, not applicable here.

Even though heat-trapping is estimated to account for only a few percent of the total energy exchange (see Trenberth et al. (2009)), it still represents a significant amount of excess heat energy that can disrupt the delicate energy balance on Earth.

An Illustrative Example

Clearly climate science is extreme complex and multidisciplinary, but it is important to understand the fundamental idea and effect of energy imbalance. We describe a very crude account, by way of an oversimplified example, to illustrate this idea.

The oceans are the largest heat sink on Earth's surface. This is because most of the water on Earth is contained in the oceans, most of Earth is covered by oceans, and—most significantly—heat capacity of water is among the largest. A slight change in ocean's surface temperature would mean a lot of thermal energy released or absorbed. Thus in our crude example, we assume Earth's surface is entirely covered with oceans. Furthermore, we assume a uniform ocean depth and temperature. Any excess energy is assumed to go into heating this body of water. These assumptions are bad for sure, but not so bad as to obscure the central idea of energy balance.

We know the energy flux, I, from the sun at the equator is $I = 1300$ W/m^2. If we assume sunlight over roughly the cross section of Earth $A = \pi R^2$ with R being the radius of Earth, the amount of heat from the sun over a time interval Δt is

$$Q = IA\Delta t. \tag{8.14}$$

The mass of water covering Earth's area $(4A)$ to depth H is given by

$$M = 4AH\rho, \tag{8.15}$$

with $\rho = 1000$ kg/m^3 being the density of water.

Let η be the fraction of the heat that is trapped, so the excess heat is ηQ. We shall call η the excess coefficient. Given the heat capacity of water to be $C = 4182$ J/kg°C, and with Eqs. (8.14) and (8.15), the change in the water temperature due to the excess heat is

$$\Delta T = \frac{\eta Q}{MC} = \frac{\eta I \Delta t}{4H\rho C}. \tag{8.16}$$

Note that the area A is canceled out. This is the final result of the example.

The result of our example Eq. (8.16) tells us that, unless the excess coefficient η is zero, a linear increase of temperature in time is predicted. Figure 8.10 shows the results from Program A.6.6 with an adjustable η. The depth of ocean is assumed to be $H = 1000$ m. At $\eta = 0.01$, one percent of the solar incident heat is trapped, and the rise in temperature over 100 years is 1.2 °C.

Even with the unrealistic assumptions in our example, a linear increase in temperature is also unlikely. We know with global warming, ice caps shrink or disappear altogether.

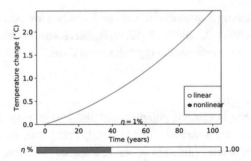

Fig. 8.10 Change of water-covered Earth surface temperature as a function of time in the oversimplified example (see text). The depth of water is assumed to be $H = 1000$ m. The adjustable excess coefficient is $\eta = 0.01$, meaning one percent of incoming solar energy is trapped

This will cause a secondary effect of worsened trapping because less sunlight is reflected back into space. There is also expected release of methane from the exposed and warming ground. There are other secondary effects. The ocean is not only a heat sink, it is also a significant CO_2 sink. With increased sea temperature, more CO_2 will be released into the atmosphere, increasing excess heat trapping. Destruction of forest, another carbon sink, caused by extreme weather and wild fire again increases greenhouse gases. These secondary effects would cause nonlinear, even runaway, temperature change.

We simulate the lowest-order nonlinear effects in Program A.6.6 with a quadratic increase by an arbitrary factor over one century (chosen to be $2 + c\eta$ in Fig. 8.10, with some factor $c \sim [0, 10]$). It is selectable in the simulation. Compared to the linear increase, a quadratic effect would double the value over a century, or 2.5 °C at $\eta = 0.01$.

Given the crude and oversimplified aspects of our example, the quantitative numbers are not to be taken too seriously (also see Exercise 8.9). However, the trend does illustrate the fundamental idea and effect of energy balance. So long as there is excess heat, even a small η, global warming is sure to continue. Even if we could magically stop the excess heat today, it would take time on the order of decades for the ocean to slowly release its heat. However, the adage that it is never too late is still true.

8.6   Exercises

8.1. Characterizing Brownian motion. We study the characteristic behaviors of position and velocity squared over different time scales in Brownian motion.

    (a) Calculate the average position and velocity squared, $\langle \mathbf{r}^2 \rangle$ and $\langle \mathbf{v}^2 \rangle$ as a function of time. The position and velocity are already stored in Numpy arrays in Program A.6.2, so the square of them may be efficiently done using vector operations with Numpy as follows,

```
for each iteration:
    call move()
    r_sqaured = np.sum(r*r)
    v_sqaured = np.sum(v*v)
    store both
```

Effectively, each statement computes the sum of vector components squared for all N particles. For instance, the position squared is $\langle \mathbf{r}^2 \rangle = \sum_i (x_i^2 + y_i^2)$ and similarly for $\langle \mathbf{v}^2 \rangle$. We could also divide these by N to obtain the averages per particle, but it is not necessary because it is a constant scale factor and we are working with arbitrary units here.

Plot the results on a log-log scale as in Fig. 8.3. How would the graph change if we used averages per particle (instead of sum over all particles)?

Looking at the $\langle \mathbf{r}^2 \rangle$ curve, how would you identify small, intermediate, and large time regions? Comment on the trends over these regions.

*(b) Fit $\langle \mathbf{r}^2 \rangle$ and $\langle \mathbf{v}^2 \rangle$ over the small and large time regions. Does a linear fit seem reasonable? If so, assume a relation $y = a + bx$ with $y = \log\langle \mathbf{r}^2 \rangle$ and $x = \log t$, and find the best-fit b in each region (see Sect. 4.5 and Program A.2.1).

Draw the fitted line over the simulation results. How do they match up visually? Do the same for $\langle \mathbf{v}^2 \rangle$.

(c) Hypothesize how the drag coefficient b and the impulse $|\Delta \mathbf{v}|$ will affect $\langle \mathbf{r}^2 \rangle$ and $\langle \mathbf{v}^2 \rangle$ in their general shape including slopes and the overall shift.

Run the simulation, separately increasing b and $|\Delta \mathbf{v}|$ by factors of two, four, etc. Compare the results with your hypothesis, and discuss the changes observed in the results. What is the main effect of increasing b or $|\Delta \mathbf{v}|$? Explain.

8.2. A game of chance. Let us play a game of chance (for a change). The rule of the game is simple. Everyone starts out with the same amount of money, say 20 pennies. In each turn, roll the dice twice to randomly choose two players. The first player chosen is to give one penny to the second player. In case the former has run out of money, skip the turn and start another one. Repeat the turns many times.

After many turns, predict what is the most likely outcome:

(a) Everyone has more or less the same amount of money
(b) A few players will end up rich and most will end up poor
(c) About half of the players will have more and the other half less
(d) Indeterminate

What thermal model does the game remind you of? Can you simulate the game to find out the correct answer? (Who knew physics could be this phun!)

8.3. Boltzmann distribution.

Calculate the Boltzmann distribution and fit it to obtain the temperature of an Einstein solid.

(a) Modify Program A.6.4 to calculate the Boltzmann distribution. The probabilities are readily returned by calling `entropy` when the solid is in equilibrium. To ensure that equilibrium is reached, let the solid cells interact for at least 10 times per cell on average, such as

```
exchange(10*N)
s, pn = entropy()
```

Graph the probability distribution P_n on a semilog scale as in Fig. 8.5.

To reduce scatter, try to average the probabilities over several samplings, allowing for some interactions in between sampling. Also, better statistics may be obtained if you increase the size of the solid $N = 1000$ or 2000, for example, but keep the average energy the same (initial energy per cell).

*(b) Fit the above probability distribution to Eq. (8.8). It is best to do a linear fit on semilog scale similar to Exercise 8.1b as $y = a - \beta n$ with $y = \log P_n$. Compare fitted β with theoretical value $\ln 2$. Graph the fitted curve with the actual distribution, normalize both at $n = 0$.

If the average energy of the system is increased, predict how the Boltzmann distribution will change. Will β be larger or smaller? Test it by doubling the average in the initialization of the solid with

```
solid = [2]*N
```

Generate the Boltzmann distribution, and repeat the fit for β. Compare it with the expected value $\ln(3/2)$.

8.4. Arrow of time. Entropy in an isolated system tends to increase, which can be viewed as an arrow of time. But how hard could it be to design a time machine to go back in time? Let us estimate it from change of entropy.

An equivalent definition of entropy to Eq. (8.9), the original one proposed by Ludwig Boltzmann himself and written on his tombstone in Vienna, is $S = k \log W$, where W (modern notation Ω) is the number of equally accessible configurations of a state (also known as the multiplicity). We can express Ω in terms of entropy as $\Omega = \exp(S/k)$.

The ratio $\Omega_f/\Omega_i = \exp(\Delta S/k)$, with $\Delta S = S_f - S_i$, gives the relative probability of finding state f to state i.

From the entropy of an Einstein solid shown in Fig. 8.6, find ΔS between the final equilibrium state and the initial uniform state. Make sure to multiply it by the number of atoms N. What is the relative probability? How likely is to go back to the initial state?

8.5. The Ising model. Investigate certain properties of the Ising model including energy and magnetization as a function of temperature. Use as template the Monte Carlo simulation, Program A.6.5.

(a) Run Program A.6.5 with the default settings. Select temperature, kT, with the slider at low, intermediate, and high values. At each point, give enough time waiting for the spin flipping and energy fluctuating as the system settles into equilibrium. Describe the spin alignment and its relationship to the total energy. Why does the energy remain negative? Note the fluctuation in energy at all but the lowest temperatures.

(b) Graph the energy as a function of temperature. The fluctuation seen above is due to the spin chain being very small ($N = 32$). To reduce fluctuation, increase the chain size to $N = 400$ (or higher), and take average over many Monte Carlo samplings, M_c. As a rule, M_c should be $M_c \geq fN$ such that each spin on average has $f \gg 1$ chances to flip to ensure equilibrium conditions. This is especially true at low temperatures where system changes slowly. The method can be summarized as

```
def energy(E, Mc):
    sum = 0
    for iterations < Mc:   # sample energy
        E = flip(E)
        sum = sum + E
    Eavg = sum/Mc          # energy per spin
    return Eavg


kT = 0.1
N = 400
f = 40
Mc = f*N
while kT < 100:
    for iterations < Mc:   # wait for equilibrium
        E = flip(E)
    Eavg = energy(E, Mc)
    append kT and Eavg
    kT = 1.3*kT
```

In the pseudo code, temperature is increased by a constant factor as opposed to a constant increment. This way, we can space out the temperature on the semilog scale efficiently.

Plot the energy vs. temperature on the semilog scale in temperature. Compare with results shown in Fig. 8.8. What temperature range has the most scatter? Why?

(c) Calculate and graph magnetization (average spin) defined as $M = \sum_i s_i/N = s/N$. Even though we could perform the sum each time, it would be inefficient. Rather, incrementally update M in the same way as updating energy. If the current net spin is s, a spin flip $s_i \to -s_i$ means $s \to s - 2s_i$. Modify the flip function as

```
def flip(E, s):
    ......
    if flip accepted:
        s = s - 2*spin[n]
    ......
    return E, s
```

Now sample s (and M) in the same way as energy (part Exercise 8.5b) for $kT = [0.1, 100]$. Graph M vs. kT on the semilog scale in temperature. Discuss the behavior of magnetization at large and small temperatures. Why is there wilder fluctuations from low to intermediate temperature? Try strategies to reduce them.

*(d) Heat capacity is defined as $C = \Delta E/\Delta T$, the energy needed to raise a unit of temperature. It is also the slope of E vs. T curve. Based on this definition and examining the shape of the energy-temperature curve in part Exercise 8.5b, sketch the C vs. T curve you expect.

It would seem that we could calculate the heat capacity by taking the numerical derivative of E with respect to T using the generated data. But this does not work because of the fluctuations inherent in Monte Carlo data of a thermal system, making numerical derivatives useless.

Instead, we use the theoretical heat capacity of the Ising model given by $C = Nk(\epsilon/kT)^2/\cosh^2(\epsilon/kT)$. Plot this result and compare with your predictions. Explain why C approaches zero at both small and large temperatures.

*8.6. Measuring entropy. We can calculate entropy from two equivalent definitions in Eq. (8.9) and Exercise 8.4. As a physical observable, even if we cannot calculate the entropy of a system, there is a way of measuring it. The change of entropy can be measured with $\Delta S = Q/T$ where Q is the heat energy absorbed by the system at constant temperature T.[4] If the temperature changes during the process, then divide it into smaller intervals such that $\Delta S = \sum_i Q_i/T_i$ where T_i is approximately constant

---

[4] Heat energy is not a state variable in thermal physics, unlike the energy of a system. So it is customarily written simply as Q, not ΔQ or dQ.

in interval i. If we start from $T \to 0$, a system is typically very ordered with zero (or very small) entropy, and we can regard ΔS the entropy at finite T.

We adopt this strategy to "measure" the entropy of the Ising model. We assume the change of internal energy of the Ising model is due to heat entering the system. The procedure is similar to the one in Exercise 8.5b,

```
initialize spin chain, E, kT
set Eold=E, kTold=kT
S = 0
while kT < 100:
    let system equilibrate
    sample avg energy, Eavg
    dE = Eavg−Eold
    Tavg = (kT + kTold)/2
    dS = dE/Tavg
    S = S + dS
    reseed Eold, kTold
    kT = 1.3∗kT
```

We use the average temperature for better accuracy. Graph the entropy as a function of temperature on the semilog kT scale. Discuss the shape of the curve in relation to that of the Einstein solid (Fig. 8.6).

Optionally, look up the exact entropy for the Ising model (see Schroeder (2021)) and plot it over your results to assess the accuracy.

*8.7. **Antiferromagnetic Ising model.** Antiferromagnetism refers to the propensity of spins (magnetic dipoles) to align antiparallel to each other because this is energetically favorable. To model an antiferromagnetic system, we can set the pair-wise coupling energy to be negative, $\epsilon < 0$, so the energy (Eq. (8.11)) of the antiferromagnetic Ising model (AIM) is $E = |\epsilon| \sum_i s_i s_{i+1}$. We investigate this model next.

(a) Assuming N spins, what are the lowest energy and the corresponding configuration of AIM? What about the highest energy and its configuration?

Find the change of energy ΔE if spin i is to be flipped. Make appropriate changes to Program A.6.5 to simulate AIM. Vary the temperature with the slider and observe the energy and configuration at low and high temperatures. Explain the results, particularly the results at higher temperatures. Does the energy ever reach the highest energy predicted above? Why?

(b) Calculate and graph energy as a function of temperature similar to that of Exercise 8.5b. Comment on the similarities and differences between the ferromagnetic and antiferromagnetic models.

Optionally, calculate the entropy and compare the results between the two models.

(c) Calculate magnetization M in the standard way and in a staggered way defined as $M_{st} = \sum_i (-1)^i s_i / N$. The latter is the so-called staggered magnetization (borrowed from 2D Ising models where it plays a more significant role).
Graph the results as a function of temperature. Compare M and M_{st}, and discuss the effect with the inclusion of the $(-1)^i$ factor in M_{st}.

8.8. Peak radiation of Earth. Calculate the peak wavelength of blackbody radiation of Earth with Wien's displacement law (Exercise 3.4), assuming a surface temperature of $T = 290$ K. Sometimes, instead of wavelengths, it is convenient to use wavenumbers defined as the inverse wavelength $1/\lambda$. Convert the result into a wavenumber in units of m^{-1} and cm^{-1}.

8.9. Global warming. Estimate the heat excess coefficient η for various scenarios in the global warming example (Sect. 8.5).

(a) Calculate the solar energy flux I at the equator of the Earth. In thermal physics, Stefan's law relates the energy flux, power per unit area, of a blackbody to its temperature T as $F = \sigma T^4$, with $\sigma = 5.7 \times 10^{-8}$ W/m^2/K^4.
Find the necessary parameters T, radius of the sun, and the distance to Earth. For our purpose, T is the surface temperature of the sun. Verify that the flux is $I = 1300$ W/m^2.

(b) The IPCC currently calls for limiting the global temperature increase to under 1.5°C of preindustrial level. Use Program A.6.6 to estimate the η necessary if temperature is to rise by that amount from the present to midcentury, or in 30 years. Give the η values for both the linear and nonlinear cases.

(c) In our example, we assumed the Earth is fully covered by oceans, when in fact the coverage is partial at about 70%. How will this affect the change in temperature? Modify the example and introduce a variable in Program A.6.6 to account for this partial coverage. How much do the results change in the linear and nonlinear cases? The depth of oceans is another parameter that we can adjust. Reduce it by half and double it, respectively. Comment on the results.

References

S. J. Blundell and K. M. Blundell. *Concepts in thermal physics*. (Oxford University Press, New York), 2010.

S. Fecht. How exactly does carbon dioxide cause global warming? *Columbia Climate School: State of the Planet*, 2021. URL https://news.climate.columbia.edu/2021/02/25/carbon-dioxide-cause-global-warming.

N. P. Gillett, E. Malinina, D. Kaufman, and R. Neukom. IPCC Sixth Assessment Report Data for Figure SPM.1. *IPCC Dataset*, 2021. URL http://dx.doi.org/10.5285/76cad0b4f6f141ada1c44a4ce9e7d4bd.

H. Gould and J. Tobochnik. *Statistical and thermal physics with computer applications*. (Princeton University Press, Princeton, NJ), 2nd edition, 2021.

N. Metropolis, A. W. Rosenbluth, M. N. Rosenbluth, A. H. Teller, and E. Teller. Equation of state calculations by fast computing machines. *J. Chem. Phys.*, **21**:1087–1092, 1953.

J. Miller. 24,000 years of climate change, mapped. *Physics Today*, **75**:14–16, January 2022.

NASA. Carbon dioxide latest measurement. *NASA Global Climate Change: Vital Signs of the Planet*, 2022. URL https://climate.nasa.gov/vital-signs/carbon-dioxide.

D. V. Schroeder. *An introduction to thermal physics*. (Oxford University Press, Oxford), 2021.

K. E. Trenberth, J. T. Fasullo, and J. Kiehl. Earth's global energy budget. *Bull. Am. Meteor. Soc.*, **90**:311–324, March 2009. URL https://doi.org/10.1175/2008BAMS2634.1.

Complex Systems 9

A *complex system* is a collection of interacting components that exhibit collective behavior and properties that are not apparent from examination of individual components. Examples include synchronized flashing of fireflies, the human brain, and global weather patterns. While there is no universal definition of "complex" behavior, complex systems are often very sensitive to initial conditions. The *butterfly effect*—the metaphorical idea that perturbations from a butterfly's wings can influence a future tornado—is a manifestation of this sensitivity.

In this chapter, we will see how seemingly simple rules governing components lead to emergent global properties. We will see that chaotic, seemingly unpredictable systems can be completely deterministic. We will also use machine learning to deduce the simple underlying rules by observing system behavior. Computation and visualization are ideal tools to study these systems, as analytic solutions and insights are often hard to extract.

9.1 Cellular Automata

A *cellular automaton* (CA) is a discrete system consisting of a grid of cells, each existing in a finite set of states. Time is also discretized. The CA starting at time t_0 is completely specified by the states of each cell and an update rule dictating how a cell's state changes. The rule is usually based on a cell's nearest neighbors and applied simultaneously to all cells, resulting in a new configuration at time $t_0 + 1$. Cellular automata can produce a wide range of complex behavior from a simple set of rules.

9.1.1 Conway's Game of Life

Conway's *Game of Life*, named after mathematician John Conway, is a 2D binary cellular automaton where each cell is either alive or dead. The cells are commonly in a rectangular

© The Author(s), under exclusive license to Springer Nature Switzerland AG 2023   203
J. Wang and A. Wang, *Introduction to Computation in Physical Sciences*, Synthesis Lectures on Computation and Analytics,
https://doi.org/10.1007/978-3-031-17646-3_9

grid and interact with their 8 bordering neighbors (including diagonally adjacent). The
update rules loosely represent a population growth model. At the next timestep:

1. Any alive cell with more than 3 alive neighbors dies (overpopulation).
2. Any alive cell with fewer than 2 alive neighbors dies (underpopulation).
3. Any alive cell with 2 or 3 alive neighbors continues stays alive (equilibrium).
4. Any dead cell with 3 alive neighbors becomes alive (reproduction).

Despite the simplicity, many complex patterns arise. Even for a small grid starting with 3
alive cells, slightly different initial conditions yield distinct pattern types as seen in Fig. 9.1.
The first row reaches a *still life*—a configuration that will not change—at $t = 1$. This can be
verified by applying the rules to each cell: each living cell has 3 living neighbors (equilibrium)
and no dead cell has 3 living neighbors (no reproduction). The second row has a slightly
different initial configuration that results in a grid of all dead cells by $t = 2$. The third row
shows another initial configuration that results in a period 2 *oscillator*—a configuration that
constantly oscillates between states.

For larger grids with random initial conditions, evolution usually begins with an initial
chaotic transient phase before still lifes and oscillators emerge. Snapshots are presented
in Fig. 9.2. While the transients are completely deterministic in the sense that there is no
randomness to the update rules, the state at $t = 0$ provides no intuition to its eventual
stable configuration at $t = 105$. In fact, changing a single cell can substantially alter its
time evolution (Exercise 9.2). Most grids do, however, eventually stabilize. Exercise 9.4
examines macroscopic properties and long term behavior of random grids.

Spaceships are another commonly occurring pattern type that translate themselves across
the grid. The *glider* is one example, illustrated in Fig. 9.3. The cells rearrange slightly at
each timestep. After 4 timesteps, every alive cell of the initial glider is translated one cell
right and one cell down. If left undisturbed, the glider will propagate infinitely. It can also
collide with distant cells, e.g. still lifes, effectively introducing nonlocal interactions further
extending or even creating new chaotic transients.

Fig. 9.1 Time evolution of 3
different initial conditions.
Alive cells are green and dead
cells are white

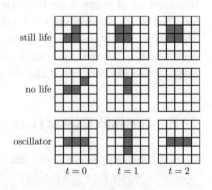

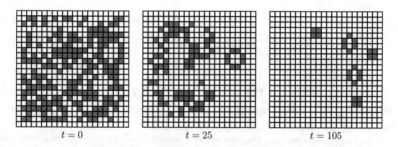

Fig. 9.2 Time evolution starting from a random grid where approximately half the cells begin alive. A stable configuration composed of 5 still lifes is reached after 105 timesteps. An intermediate still life, the right-most pattern at $t = 25$, is destroyed in the process

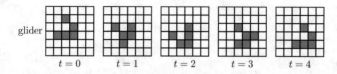

Fig. 9.3 Time evolution of a moving glider spaceship

To implement the Game of Life, we represent a grid as a numpy array taking values 1 (alive) or 0 (dead). In theory, the Game of Life can be played on an infinite grid and implemented using dynamically updating grids or even a non-grid representation, but we will impose boundary conditions instead. Figure 9.2 was generated with *constant boundary conditions* that constrain all cells on the boundary to take constant value 0, i.e. are always dead. They are the simplest to implement, but yield artificial behavior at the boundaries. To remove these edge effects, *periodic boundary conditions* can be used instead. With periodic boundaries, the game is played on the surface of a torus. We will discuss implementation with constant boundaries; periodic boundaries are a straightforward extension explored in Exercise 9.3.

The main function we need takes an initial grid, applies the Game of Life rules, and returns the updated grid. The simplest method loops through all cells, except the constant boundary cells, and stores the updated values in the next grid.

```
In [1]: import numpy as np

        def update_grid(grid):
            n_row, n_col = grid.shape
            next_grid = np.zeros(grid.shape, dtype=int)
            for i in range(1, n_row-1):
                for j in range(1, n_col-1):
                    cell = grid[i, j]
                    neighborhood = grid[i-1:i+2, j-1:j+2]
                    n_alive_neighbors = np.sum(neighborhood) - cell
                    next_grid[i, j] = update_cell(cell, n_alive_neighbors)
            return next_grid
```

The update rules are encoded in `update_cell()`, which can be concisely expressed as

```
In [2]: def update_cell(cell, n_alive_neighbors):
            if cell:
                return int(2 <= n_alive_neighbors <= 3)
            else:
                return int(n_alive_neighbors == 3)
```

Given an initial grid, we can sequentially apply `update_grid()` to determine its time evolution. A random 20×20 grid can be initialized as

```
In [3]: grid = np.random.randint(low=0, high=2, size=(20, 20))
        grid = np.pad(grid, pad_width=1, mode='constant', constant_values=0)
```

The padding adds constant boundaries. All that remains is a way to visualize the grid and its time evolution. We do so using `matplotlib`, in particular its `animation` submodule.

```
In [4]: import matplotlib.pyplot as plt
        import matplotlib.animation as animation
        from matplotlib import colors

        def play(grid, T=100, show=True):
            fig, ax = plt.subplots(figsize=(5, 5))
            ax.set(xticks=[], yticks=[], aspect='equal')
            ims = []
            for _ in range(T + 1):
                im = ax.pcolormesh(grid, edgecolor='k', linewidth=.01, animated=True,
                                   cmap=colors.ListedColormap(['white', 'green']))
                ims.append([im])
                grid = update_grid(grid)
            anim = animation.ArtistAnimation(fig, ims, interval=100, blit=True)
            plt.show() if show else plt.close()
            return anim
```

The main calculations take place in the `for` loop. An image `im` is stored at each timestep using `pcolormesh()`, which generates a black mesh coloring alive cells as green and dead cells as white. The images are stored in the list `ims` which must be a list of lists, hence the appending of `[im]`. The images are then stitched together using `ArtistAnimation()` and returned in the object `anim`. By default, the animation is played when using a command-line interface. When using Python interactively, e.g. in Jupyter Notebook, we must use one of `anim`'s built-in methods to convert the animation. The full sequence of Fig. 9.2 can be viewed interactively by running

```
In [5]: np.random.seed(1)
        grid = np.random.randint(low=0, high=2, size=(20, 20))
        grid = np.pad(grid, pad_width=1, mode='constant', constant_values=0)
        anim = play(grid, T=110, show=False)

        from IPython.display import HTML
        HTML(anim.to_jshtml())
```

The animation can also be saved using the `anim`'s `save()` method, e.g. `anim.save('life.mp4')`.

9.1.2   Traffic Flows

Cellular automata can be used to model complex behavior and thus help understand it from a simple set of rules. One example is the Nagel–Schreckenberg model for traffic (see Nagel and Schreckenberg 1992). It is a one-dimensional CA for a single-lane road that can reproduce traffic jams. We will use periodic boundary conditions, corresponding to a circular road. Open boundaries yield similar qualitative results, but require a few more modeling steps (Exercise 9.7). The road is represented by an array of cells that can be empty or occupied by a car with integer velocity v ranging from 0 to some speed limit v_{max}. Cars are updated simultaneously in parallel. Their subsequent positions and velocities are updated according to the following ordered rules:

1. If $v < v_{max}$, increase $v \to v + 1$ (acceleration up to speed limit).
2. If there are only $d < v$ empty cells before the next car, decrease $v \to d$ (avoid collision).
3. If $v > 0$, randomly decrease $v \to v - 1$ with probability p_{slow} (random deceleration).
4. Increase car position by v (motion).

Though the model is 1D, it is more complex than the Game of Life in two ways. First, each cell has $v_{max} + 2$ possible states. Second, and more importantly, the model has a random component. The random component mimics human behavior and turns out to be essential for traffic jams. Without it, cars quickly reach a repeating steady state configuration (Exercise 9.6). As the state of a road cannot be represented by just an array, we will implement a Road class to store associated data.

```
In [6]:  class Road:
             def __init__(self, length, density, p_slow=.1, v_max=5):
                 n_car = int(np.round(length * density))
                 xs = sorted(np.random.choice(length, size=n_car, replace=False))
                 vs = np.zeros(n_car, dtype=int)
                 self.cars = [(xs[i], vs[i]) for i in range(n_car)]
                 self.n_car, self.length, self.density = n_car, length, density
                 self.p_slow, self.v_max = p_slow, v_max

             def update(self):
                 updated_cars = []
                 for i in range(self.n_car):
                     x, v = self.cars[i]
                     front_x, front_v = self.cars[(i+1) % self.n_car]
                     dist = (front_x - 1 - x) % self.length
                     if v < self.v_max: v += 1
                     if dist < v: v = dist
                     if v > 0 and np.random.rand() < self.p_slow: v -= 1
                     updated_cars.append(((x + v) % self.length, v))
                 self.cars = updated_cars
```

All Road instances take two required parameters: the length of the road (in cells) and the density of cars occupying it. For the optional parameters, default values correspond to the realistic case $p_{slow} = 0.1$ and $v_{max} = 5$. With this speed limit, each timestep is on the time scale $\mathcal{O}(1\,\mathrm{s})$, comparable to a human driver's reaction time.

The __init__() method is called on every Road instance and initializes a list of car position-velocity pairs in the cars attribute as well as road-specific constants in the other attributes. The cars are randomly placed at rest. Since in one-dimension cars cannot pass, a given car always checks the same car in front of it for collision detection. Sorting cars by position preserves the ordering for all future updates. Note that we have used a *list comprehension*, which is simply notational shorthand for

```
self.cars = []
for i in range(n_car):
    self.cars.append((xs[i], vs[i]))
```

The update() method implements the above four rules. Updated position-velocity pairs are stored in a new list so that cars are updated simultaneous in parallel. The % operator conveniently implements periodic boundaries by automatically wrapping around to the other side without case by case if statements. The following creates an instance of Road and updates the cars once:

```
In [7]:  road = Road(length=100, density=.05)
         road.update()
```

Time evolution of a road is best illustrated with *spacetime plots*; examples at two different densities are shown in Fig. 9.4. The state of the road is plotted at each time. Cells occupied by cars are shaded, where lighter shading corresponds to faster speeds. Empty cells are white.

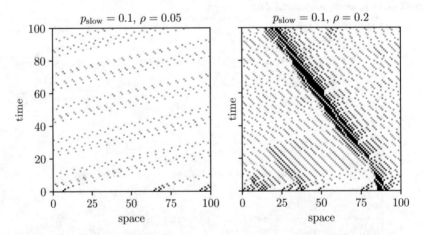

Fig. 9.4 Spacetime plot of a sparse and dense road of cars that start at rest with speed limit $v_{max} = 5$. Shaded cells represent car locations at each time; the lighter the shading, the faster the car. The sparse road flows near the speed limit while the dense road creates traffic jams

Consider the sparse road, $\rho = 0.05$. At $t = 0$, there are five black squares representing cars that start at rest. At time $t = 1$, the left-most car's velocity becomes 1 and thus its position increases by 1. Similarly at time $t = 2$, its velocity becomes 2 and its position increases by 2. This continues until it reaches $v_{max} = 5$ at $t = 5$. The left-most car's time evolution is simple as there are no nearby cars in front and no initial random slow downs. The other four cars eventually settle into similar trajectories that appear as approximate parallel lines with slope $1/v_{max}$. This is because at this low density, the cars are able to flow freely at the speed limit. At each time update, each car translates in space by approximately 5 cells.[1]

In contrast in the dense road, $\rho = 0.2$, we observe a dark band of stationary cars occupying nearly half the cars, i.e. a traffic jam. The traffic jam propagates backward in space. This is because before the jam, the flow of cars is actually quite high whereas cars in the front of the jam must slowly accelerate. Close examination of the jam around $t = 30$ shows it is caused by random slow downs of just a few cars. Though p_{slow} is small, the effects propagate to all cars in the neighborhood behind. When the density of cars is small, the effects are less dramatic and no traffic jams appear.

The onset of traffic jams can be viewed as a phase transition tuned by the road density. To determine the critical density ρ_* separating free flow from jamming, we can measure the average flow of cars passing through a given point. We expect a sharp transition at ρ_*. The flow can be measured by placing a detector at some equilibrium point on the road that counts the number of cars N that pass in T timesteps. The time averaged flow is N/T.

Figure 9.5 plots flows as a function of density for two different p_{slow} values. The two plots are qualitatively similar and display three regimes. First, for low densities, the cars are in free flow so the average flow increases linearly with density—there is enough space that each car can safely accelerate to v_{max}. Second, there is a sharp transition at $\rho_* = 0.17$ and $\rho_* = 0.11$ for $p_{slow} = 0.1$ and $p_{slow} = 0.5$, respectively. This regime marks the onset of traffic jams where the flow dramatically decreases. The sharp transition is clearer for $p_{slow} = 0.5$. This regime is very short lived and due to the discrete nature, we cannot improve our density resolution. Finally, once traffic jams appear, further increasing the density decreases the flow due to increased congestion, but not as dramatically as the second regime. Though short lived, the dramatic decrease in flow in the second regime is crucial evidence for the onset of traffic jams. Indeed it is absent in the deterministic case $p_{slow} = 0$ where no traffic jams appear (Exercise 9.6).

The Nagel–Schreckenberg model can be extended to multi-lane traffic by adding lane-switching rules. However, as the simpler one-lane model already captures essential features of traffic jams, we will not pursue those and other generalizations.

---

[1] Approximate because of random deceleration.

Fig. 9.5 Average flow over 10^5 timesteps as a function of density. Flow is only measured after 10^3 timesteps to avoid initial transients. The roads are 100 cells long and $v_{max} = 5$

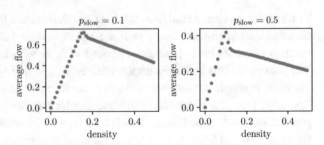

9.2   Chaos

We tend to associate chaos in the colloquial sense with disorder, irregularity, and randomness. However in the mathematical sense, chaos is completely deterministic. The apparent contradiction in definitions is resolved by a defining feature of chaos—extreme sensitivity to initial conditions. This sensitivity gives chaotic systems the semblance of randomness, as even a small rounding error can lead to wildly differing results over time. Weather systems are a prime example of chaos. They are composed of many components that interact in a nonlinear fashion, and even small disturbances can lead to dramatic changes over time. This is what makes long term weather forecasts more than a week away difficult, even though day to day forecasts are reliable.

Chaos can appear in simple systems as long as nonlinearity is present. We will consider two 1D systems that exhibit chaotic behavior and features that are universal to all chaotic systems. We will also introduce analytical and computational tools to understand chaos. For models of more complex systems like weather, see Strogatz (1994), Wang (2016).

9.2.1   Logistic Map

The *logistic map* is one of the simplest systems that exhibits chaotic behavior. It takes a value $x_t \in [0, 1]$ at time t and maps it at the next timestep to value

$$x_{t+1} = 4rx_t(1 - x_t), \quad r \in [0, 1], \; x_t \in [0, 1]. \quad \text{(Logistic map definition)} \quad (9.1)$$

To avoid confusion between a value the system takes and the function that maps x_t to x_{t+1}, it is useful to introduce the *map function* $f(x) = 4rx(1 - x)$ so that Eq. (9.1) is equivalent to $x_{t+1} = f(x_t)$. The restricted domain of r ensures that $x_t \in [0, 1]$ for all t. Time is discretized, as was the case in cellular automata, but now x_t is continuous. The logistic map can be interpreted as a simple population model. The map function is a parabola with maximum at $x = \frac{1}{2}$. For populations $x < \frac{1}{2}$, $f(x)$ is an increasing function of x, capturing increasing reproduction rates. In contrast for $x > \frac{1}{2}$, $f(x)$ is a decreasing function of x, capturing increasing overcrowding and starvation rates. The exact rates are controlled by parameter r. We will not focus on model interpretation, but rather on model behavior.

The logistic map is straightforward to implement. The map function is

```
In [8]: def logistic_map(x, r):
            return 4*r * x * (1-x)
```

We can learn about the temporal behavior by applying the map multiple times, starting with an initial value x_0 and plotting x_t over time. A *cobweb plot* is an alternative representation that explicitly visualizes the map's role and allows for inference into evolution with different initial conditions. An example of both is shown in Fig. 9.6 with $(x_0, r) = (0.1, 0.7)$. The left plot shows the x_t trajectory over time. Starting at $x_0 = 0.1$ at $t = 0$, Eq. (9.1) maps x_0 to $x_1 = 4 \times 0.7 \times 0.1 \times 0.9 = 0.252$ at $t = 1$. This process continues and we see x_t grow until $t = 5$, after which x_t seemingly approaches an equilibrium point (also known as fixed point) with less under- and overshoot over time. The right cobweb plot tells a similar story; it is constructed as follows:

1. Plot the map function $f(x_t)$ and the line $x_{t+1} = x_t$. Start at coordinate $(x_0, 0)$.
2. Advance vertically to $f(x_t)$, which gives the coordinate of the updated value x_{t+1} for any t.
3. Advance horizontally to the line $x_{t+1} = x_t$. This projects x_{t+1} on to x_t and allows for repeated iteration of the previous step.
4. Repeat steps 2 and 3 for T timesteps and mark the endpoint.

The cobweb representation allows us to quickly conclude there are two equilibrium points. An equilibrium point is like a still life—once reached, the system will always stay there. Mathematically, this means $f(x_t) = x_t$. Geometrically, this is the intersection between the map and the line in the cobweb plot, which occurs at two (equilibrium) points: $x_* = 0$ and

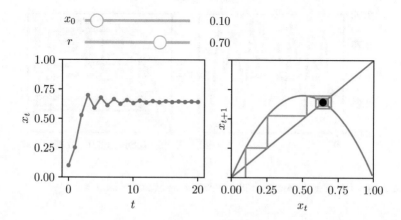

Fig. 9.6 A snapshot of an interactive slider plot showing evolution toward an equilibrium point

$x_* = \frac{9}{14}$. The exact values of the equilibrium points can be found by solving $f(x_t) = x_t$, or equivalently $f(x_t) - x_t = 0$. The latter form is convenient for solving via sympy:

```
In [9]:   import sympy as sp
          x, r = sp.symbols('x r')
          sp.solveset(logistic_map(x, r) - x, x)
```

Out[9]: $\left\{ 0, \frac{4r-1}{4r} \right\}$

Note that the second equilibrium point is negative for $r < \frac{1}{4}$, implying $x_* = 0$ is the only equilibrium point there. Geometrically, the parabolic map never exceeds the line in the cobweb plot. Increasing r past $\frac{1}{4}$ allows it to and creates a second equilibrium.

9.2.2   Stability of Equilibrium Points

Knowing the exact values of x_* does not necessarily tell us the behavior of the system. For instance in Fig. 9.6, the equilibrium point farther away from the initial value x_0 is eventually reached. Indeed tracing out a cobweb for any x_0 near $x_* = 0$ will lead away from it. We classify equilibrium points like this as *unstable*—unless $x_0 = 0$ exactly, the system will never reach it. In contrast, any web near $x_* = (4r - 1)/(4r)$ spirals toward it; it is *stable*. Stability is not a permanent property. Increasing r further creates a system with two unstable equilibrium points, shown in Fig. 9.7. Moreover, there is no longer any structure as the system wildly fluctuates in a seemingly random manner, a trademark of chaos. As Figs. 9.6 and 9.7 are just two snapshots in a wide range of possibilities, it is best if you run the following in an interactive notebook to generate your own figures:

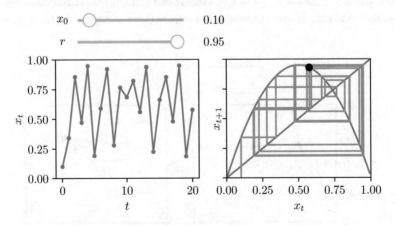

Fig. 9.7 A snapshot of an interactive slider plot showing chaotic evolution

In [10]:
```python
import ipywidgets

def cobweb_coords(x0, r, T=20):
    x1 = logistic_map(x0, r)
    coords = [(x0, 0), (x0, x1)]
    for _ in range(T):
        x0, x1 = x1, logistic_map(x1, r)
        coords.extend([(x0, x0), (x0, x1)])
    return coords

ipywidgets.interact(x0=ipywidgets.FloatSlider(min=0, max=1, step=.01),
                    r=ipywidgets.FloatSlider(min=0, max=1, step=.01))
def evolution_plots(x0, r):
    web_xs, web_ys = list(zip(*cobweb_coords(x0, r)))
    traj = [x for i, x in enumerate(web_xs) if i % 2 == 0]
    xs = np.arange(0, 1, .01)

    fig, axs = plt.subplots(nrows=1, ncols=2, figsize=(7, 3), sharey=True)
    axs[0].plot(traj, linestyle='-', marker='.')
    axs[0].set(xlabel='$t$', ylabel='$x_t$', ylim=[0, 1])
    axs[1].plot([0, 1], [0, 1], color='C0')
    axs[1].plot(xs, logistic_map(xs, r), color='C0')
    axs[1].plot(web_xs, web_ys, color='C1')
    axs[1].plot(web_xs[-1], web_ys[-1], color='black', marker='o')
    axs[1].set(xlabel='$x_t$', ylabel='$x_{t+1}$', xlim=[0, 1])
```

Given an initial value x_0 and parameter r, cobweb_coords() returns a list of cobweb coordinates for T timesteps. The ipywidgets.interact() decorator enables interactive sliders immediately after the plotting function is run. evolution_plots() unzips the list of coordinate pairs into separate lists of x and y coordinates, extracts the x_t trajectory, and makes the trajectory and cobweb plots. Choose any value of x_0, slowly increase r from 0 to 1, and observe the change in stabilities of the equilibrium points.

We can analyze equilibrium point stabilities by considering small perturbations around them. Suppose at time t the system is at x_t in the neighborhood of an equilibrium point x_*, a distance

$$|\epsilon_t| \equiv |x_t - x_*| \tag{9.2}$$

away. At the next timestep, a stable equilibrium point must satisfy

$$|\epsilon_{t+1}| < |\epsilon_t|, \tag{9.3}$$

i.e. the system approaches the equilibrium point. We can relate the two distances with the map function:

$$|\epsilon_{t+1}| = |x_{t+1} - x_*| = |f(x_t) - x_*| = |f(x_* + \epsilon_t) - x_*|. \tag{9.4}$$

For small ϵ_t, it is sufficient to expand to leading order: $f(x_* + \epsilon_t) = f(x_*) + f'(x_*)\epsilon_t + \mathcal{O}(\epsilon_t^2)$. Since $f(x_*) = x_*$, by definition of an equilibrium point, substitution into Eq. (9.4) yields

$$|\epsilon_{t+1}| = |f'(x_*)\epsilon_t| = |f'(x_*)||\epsilon_t|. \tag{9.5}$$

It follows from Eq. (9.3) that the stable equilibrium condition is

$$\left| f'(x_*) \right| < 1. \quad \text{(Stable equilibrium)} \tag{9.6}$$

Similarly, the unstable equilibrium condition is $\left| f'(x_*) \right| > 1$. The borderline case $\left| f'(x_*) \right| = 1$ requires analysis of higher order terms. We can again use `sympy` to evaluate the stability condition for the logistic map's equilibrium points, displaying as a dictionary of equilibrium point and stability condition pairs:

```
In [11]:  sp.init_printing(use_latex='mathjax')   # For pretty printing
          solset = sp.solveset(logistic_map(x, r) - x, x)
          for sol in solset:
              display({sol: logistic_map(x, r).diff(x).subs(x, sol).simplify()})
```

Out[11]: $\{0 : 4r\}$

$$\left\{ \frac{4r-1}{4r} : 2 - 4r \right\}$$

This tell us that $x_* = 0$ becomes unstable for $r > \frac{1}{4}$ and $x_* = (4r-1)/(4r)$ becomes unstable for $r > \frac{3}{4}$, which you can verify using the interactive plots.

9.2.3   Onset of Chaos

Loss of stability alone does not guarantee the chaotic behavior seen in Fig. 9.7. Trajectories could be periodic, for example. To investigate the transition to chaos, it is useful visualize long term behavior as a function of r. We can do this by applying the map $1,000$ times and then recording the next 100 values of x in small increments of r, say $dr = 0.001$. Figure 9.8(left) contains the resultant *bifurcation diagram*, generated with

```
In [12]:  def bifurcation_plot(rs=np.arange(0, 1, .001), n_toss=1000, n_keep=100):
              fig, ax = plt.subplots()
              ax.set(xlabel='$r$', ylabel='$x$')
              for r in rs:
                  x, xs_keep = 0.5, []
                  for _ in range(n_toss): x = logistic_map(x, r)
                  for _ in range(n_keep):
                      x = logistic_map(x, r)
                      xs_keep.append(x)
                  ax.plot([r]*n_keep, xs_keep, 'C0,')
```

We plot in batches for efficiency. For simplicity, we always use $x_0 = \frac{1}{2}$, but the results are robust to different initial starts.[2] This is because in the presence of stable equilibrium points, we expect trajectories to converge to that point and in the presence of chaos, we expect no regularity anyway. Indeed this is what we observe. For $r < \frac{1}{4}$, all trajectories converge to the equilibrium point $x_* = 0$. This point becomes unstable and a new equilibrium point is created at $r = \frac{1}{4}$. The bifurcation diagram confirms our stability analysis as we see that equilibrium point is only stable for $r \in (\frac{1}{4}, \frac{3}{4})$.

The interesting dynamics occur for $r > \frac{3}{4}$ when we observe a bifurcation as the stable equilibrium point forks into two branches, magnified in Fig. 9.8 (right). This corresponds to

---

[2] If you prefer, you could use random starts with a fixed seed for reproducibility.

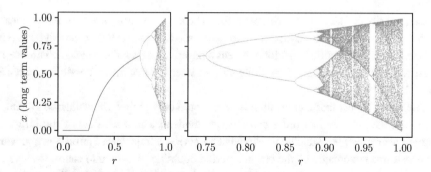

Fig. 9.8 Bifurcation diagram of the logistic map, showing the long term behavior as a function of r. The right plot magnifies the period doubling and chaotic regions

a trajectory with period 2, i.e. one that oscillates between two stable points. In the magnified view, we see that each of these branches forks into two more near $r_4 = 0.862$ and again near $r_8 = 0.886$, where the subscripts indicate the period after the branching. One can find period 16, 32, ... trajectories computationally (Exercise 9.8) with rapidly shrinking distances between doublings as we have already seen: $[0.862 - 0.75, 0.886 - 0.862, \ldots] = [0.112, 0.024, \ldots]$. Though the proof is out of the scope of this book, the ratio of distances, e.g. $0.112/0.024 \approx 4.67$, approaches a constant value $\delta = 4.669...$ known as *Feigenbaum's constant*. This means an *infinite* number of period doublings occur in a short window, as the geometric series $0.75 + 0.112 \sum_{i=0}^{\infty} \delta^{-i} < 0.9$, and chaos emerges.

The chaotic regime begins just before $r = 0.9$ and can be seen as a dense strip of points in the bifurcation diagram, Fig. 9.8, as it represents an infinite set of points. We mentioned earlier that chaotic systems are extremely sensitive to initial conditions. To show this is the case in the logistic map, Fig. 9.9 (left) plots two trajectories separated initially by $\Delta x_0 = 0.0001$. After just 20 timesteps, the trajectories have diverged completely. The absolute difference in trajectories in the right plot shows that they diverge exponentially fast, indicated

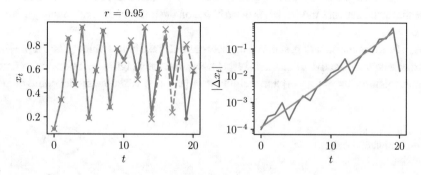

Fig. 9.9 Two trajectories that quickly diverge, despite differing initially by only $\Delta x_0 = 0.0001$, showing sensitivity to initial conditions in a chaotic regime of the logistic map

by the linearity on the semilogarithmic scale. The slope represents a quantitative measure of chaoticity, known as the Lyapunov exponent (see Strogatz 1994). A positive Lyapunov exponent indicates chaotic behavior, whereas a negative one regular behavior. The steep dip at the end is just an artifact of the finite system size: the maximum possible difference is $\Delta x = 1$.

One might expect that increasing r further would only yield a more chaotic system, but this is not the case. In fact order can emerge, evidenced in a window of the bifurcation diagram near $r = 0.96$, which corresponds to a period 3 trajectory. Period 3 is noteworthy because it was not created by the original period doubling cascade into chaos. Remarkably, zooming in on any of the three branches yields a bifurcation diagram that almost looks like an identical copy (Exercise 9.9). This self similarity exists in other regions and is another feature of chaotic systems in general. The period doubling cascade to chaos also is not unique to the logistic map; more examples are explored in Exercise 9.10.

9.3 Neural Networks

Neural networks have excelled in many tasks in recent years as computational capabilities have evolved. These tasks include image recognition, speech detection and language translation, competition in games like chess or StarCraft, and many more, including tasks in physics and engineering. In this section, we will study the foundational structure of a neural network and how they learn from data. We will then train a neural network to learn the rules of the Game of Life by observing how grids evolve.

9.3.1 A Simple Network

Consider a network of just three neurons, schematically shown in Fig. 9.10. The network consists of two input neurons which take values x_1 and x_2. The output neuron then uses the input values, weighted by w_1 and w_2, to output value y. The exact function used to output y depends on the network model, but is often based on the linear combination $w_1 x_1 + w_2 x_2 = \mathbf{w} \cdot \mathbf{x}$.

As a further simplification, suppose the neurons only take binary values 1 or 0 representing whether a neuron is active or inactive, respectively, but the weights can be any real numbers. Then given inputs $\mathbf{x} = [x_1, x_2]$ and weights $\mathbf{w} = [w_1, w_2]$, a simple model could output

Fig. 9.10 Graphical representation of a simple network

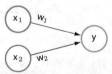

$$y = f(\mathbf{w} \cdot \mathbf{x} + b) = \begin{cases} 1 & \text{if } \mathbf{w} \cdot \mathbf{x} + b > 0, \\ 0 & \text{if } \mathbf{w} \cdot \mathbf{x} + b \leq 0, \end{cases} \qquad (9.7)$$

where b is a constant *bias* that sets the output neuron's baseline disposition to activate. $f(\cdot)$ is known as the *activation function* because it determines if a neuron activates given parameters of the network, $\mathbf{w}$ and b, and its inputs, $\mathbf{x}$. For example, a network with $b = -2.5$ and weights $\mathbf{w} = [1, 2]$ is biased against activating, but the positive weights push the output to activate when the inputs are active. In this case, $y = 1$ only when both inputs are active, i.e. it performs the logical and operation x1 and x2. A negative weight would inhibit activations (when the input is active). The size of the weight determines the relative importance of a connection—a weight of zero means the input neuron has no influence over the output's activation. This simple network can be implemented in Python as follows:

```
In [13]:  class SimpleNetwork:
              def __init__(self, w1, w2, b):
                  self.w1, self.w2, self.b = w1, w2, b

              def predict(self, x1, x2):
                  return int(self.w1*x1 + self.w2*x2 + self.b > 0)

          net = SimpleNetwork(w1=1, w2=2, b=-2.5)
          net.predict(x1=1, x2=1)
```

Out[13]: 1

It is reassuring that the simplified binary case can produce a logical operator, but the real utility comes from a neural network's ability to learn from data. In that regard, the binary model is unsatisfactory for a number of reasons, a major one being that data is often continuous. For example, suppose we are interested in determining whether an image contains or does not contain a cat. The output is binary, but the image input data are pixel intensities. It would be wasteful and inaccurate to reduce each pixel in an image to just a 1 or 0. Relaxing the binary input requirement helps, say letting $x_i \in [0, 1]$. However from a learning perspective, the step activation function in Eq. (9.7) is still too abrupt. To see this, think about a learning algorithm tasked to finding the optimal parameters. Optimal in this context means the network's output best matches the output of the actual data, e.g. the network correctly classifies the most cat images. The algorithm would like to finely tune the weights w_i and bias b with continuous feedback based on the output of the network and the data. With the step activation function, there is only feedback at the single point where $\mathbf{w} \cdot \mathbf{x} + b$ crosses zero, so fine tuning of the parameters is not possible.

A better activation function smooths out the step function and typically are *sigmoid functions* with a characteristic "S"-shaped curve. We will be using the *logistic function*[3] for classification, again based on the linear combination $z \equiv \mathbf{w} \cdot \mathbf{x} + b$:

$$f(z) = 1/(1 + e^{-z}). \qquad (9.8)$$

---

[3] The name is derived from being the solution to a continuous version of the logistic map.

Fig. 9.11 Comparison of step
and logistic activation
functions, Equations (9.7) and
(9.8)

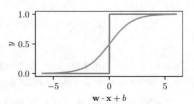

It is plotted in Fig. 9.11 in comparison to the step function. The logistic function smoothly transitions from 0 to 1, allowing for continuous feedback to obtain the optimal parameters. Furthermore if the underlying data truly did have a step-like transition, the neural network could capture that by increasing the weights and bias by a constant factor. Though y is now continuous, binary classification can be achieved by thresholding, e.g. classify as a cat if $y >$ 0.5. This provides more flexibility because we can change the threshold depending on the application. For example, increasing the threshold to $y > 0.9$ results in a more conservative classifier with fewer false positives.

9.3.2 Extending the Network

The most straightforward and practical extension of the three neuron network adds more input neurons. For example, an image with p pixels can be represented as a vector of p inputs. It is also useful to add more output neurons so we can classify images with q categories. Supposing we are interested in labeling cats and dogs, we have $q = 2$ represented by two outputs y_1 and y_2. We use y_1 to determine whether the image contains a cat and y_2 to determine whether the image contains a dog. Note that the categories do not need to be mutually exclusive.

We will consider *fully connected networks*, where each input neuron is connected to every output neuron for a total of $p \times q$ weights and q biases, illustrated in Fig. 9.12 (left). To simplify the notation, we have represented the biases as ordinary weights w_{0l} stemming from an input neuron with value $x_0 = 1$. Each output has a different set of weights and bias, so there are q output equations:

$$y_l = f(z_l), \quad l = 1, 2, \ldots, q. \tag{9.9}$$

As usual, $f(z_l)$ is the activation function based on the linear combination $z_l = \mathbf{w}_l \cdot \mathbf{x}$, where the bias has been incorporated into the weights $\mathbf{w}_l = [w_0, w_1, w_2, \ldots, w_p]$ and input data $\mathbf{x} = [1, x_1, x_2, \ldots, x_p]$.

While linearity is often a good approximation, it limits the range of patterns the neural network can learn (Exercise 9.11). To overcome this limitation, we can add intermediate layers to the network as in Fig. 9.12 (right). The r neurons in the intermediate layer use $\mathbf{x}$ as inputs with (bias and) weights $\mathbf{w}_k = [w_{0k}, w_{1k}, w_{2k}, \ldots, w_{pk}]$ to determine their outputs h_k,

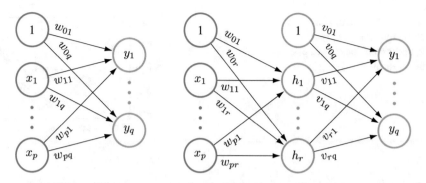

Fig. 9.12 Graphical representation of fully connected neural networks

$k = 1, 2, \ldots, r$. After appending a neuron $h_0 = 1$ to incorporate new biases, they are fed as inputs into the output layer with another set of (bias and) weights $\mathbf{v}_l = [v_{0l}, v_{1l}, v_{2l}, \ldots, v_{rl}]$. Because the h_k are not of direct interest, they are known as *hidden neurons*. The outputs are then computed from

$$h_k = g(\mathbf{w}_k \cdot \mathbf{x}), \quad k = 1, 2, \ldots, r,$$
$$y_l = f(\mathbf{v}_l \cdot \mathbf{h}), \quad l = 1, 2, \ldots, q, \tag{9.10}$$

where $g(\cdot)$ and $f(\cdot)$ are possibly different activation functions. As long as $g(\cdot)$ is a nonlinear function of the inputs, then so are the final outputs y_l. In general, there can be multiple hidden layers, each additional layer increasing the complexity of patterns the network can learn. More complexity is not always better, however, as we do not want to learn random noise in a dataset which would generalize poorly to new data. One hidden layer is a good starting point, though some experimentation on the number of layers and neurons in each layer is usually needed.

Aside from smoothness, nonlinearity is the main requirement for activation functions in the hidden layer(s). The degree of nonlinearity is not crucial, as more hidden neurons can be added for increased complexity. Indeed the simple *ramp* activation function

$$g(z) = \max(0, z) = \begin{cases} z & \text{if } z > 0, \\ 0 & \text{otherwise.} \end{cases} \tag{9.11}$$

is popular, in part due to its simplicity and thus computational efficiency. Though the derivative is undefined at $z = 0$, in practice we just choose it to be 0 or 1 with little impact on the final model. A neuron with ramp activation is referred to as a *rectified linear unit* (ReLU). Linear combinations of ReLUs can closely approximate nonlinear functions, demonstrated in Fig. 9.13. With just $r = 2$ units, the quadratic approximation is crude, but quickly improves as we add units (hidden neurons). Many variations of ReLUs exist, e.g. modifications that smooth out the discontinuous derivative. Hidden sigmoid activations are also common, but ReLUs are good enough for our purposes.

Fig. 9.13 Approximating a
quadratic function $f(x) = x^2$
using linear combinations of 2,
4, and 16 ReLUs

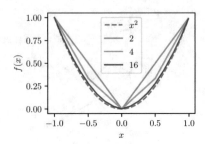

Choice of activation function in the final output layer $f(\cdot)$ depends on the application of the neural network. For classification, the logistic function works well because it is rooted in statistical theory. Indeed, its output can be interpreted as a probability. Our previous thresholding rule $y > 0.9$ can be interpreted as "classify as a cat if the predicted probability of being a cat is greater than 0.9". When the categories are mutually exclusive, the *softmax* function should be used:

$$\text{softmax}(z)_l = \frac{e^{z_l}}{\sum_{l=1}^{q} e^{z_l}}. \tag{9.12}$$

The denominator is a normalization constant, analogous to the thermodynamic partition function. When a neural network is used to approximate an unknown function, $f(\cdot)$ is commonly the identity function so the network output (with $q = 1$) is just the linear combination $y = \mathbf{v} \cdot \mathbf{h}$. The output is unbounded and nonlinearities are encoded in the hidden layers, so a large class of functions can be learned (Exercise 9.12).

9.3.3 Learning from Data

In order for a network to learn, we must *train* it using data. Suppose we have a dataset with N observations, each observation consisting of a vector of inputs $\mathbf{x}_i = [1, x_{i1}, x_{i2}, \ldots, x_{ip}]$ and a vector of outputs $\mathbf{y}_i = [y_{i1}, y_{i2}, \ldots, y_{ir}]$, $i = 1, 2, \ldots, N$. For a network with one hidden layer, as in Fig. 9.12 (right), training amounts to finding the optimal sets of weights which we represented as matrices $\mathbf{W}$ and $\mathbf{V}$, whose elements are w_{jk}, $j = 0, 1, \ldots, p$ and $k = 1, 2, \ldots, r$, and v_{kl}, $k = 0, 1, \ldots, r$ and $l = 1, 2, \ldots, q$. The kth column of $\mathbf{W}$ are the $\mathbf{w}_k$ in Eq. (9.10), and similarly the lth column of $\mathbf{V}$ are the $\mathbf{v}_l$. Note the slightly different range of k values depending on if it is the first or second index; this is because no connections are directed into the bias neuron, or equivalently $w_{j0} = 0$.

Denote the network's predicted outputs as $\widehat{\mathbf{y}}_i$, which is a function of $\mathbf{W}$ and $\mathbf{V}$. Ideally we would want $\widehat{\mathbf{y}}_i$ to closely match the data $\mathbf{y}_i$, so it is intuitively appealing to measure a network's performance based on the error $\mathbf{y}_i - \widehat{\mathbf{y}}_i$. Since positive and negative errors should be treated equally, we use the squared errors $(\mathbf{y_i} - \widehat{\mathbf{y}}_\mathbf{i}) \cdot (\mathbf{y_i} - \widehat{\mathbf{y}}_\mathbf{i}) \equiv (\mathbf{y}_i - \widehat{\mathbf{y}}_i)^2$. Furthermore, treating each data observation equally, our goal is to minimize the *loss function*[4]

$$L(\mathbf{W}, \mathbf{V}) = \frac{1}{N} \sum_{i=1}^{N} (\mathbf{y}_i - \widehat{\mathbf{y}}_i)^2, \tag{9.13}$$

sometimes referred to as the *mean squared error* (see also Sect. 4.5). The *mean* rather than *sum* of squared errors is not strictly necessary, but it is useful in practice for training, which we will see later.

In principle, the task is simple. We just minimize the loss function, Eq. (9.13), with respect to the weights using our favorite optimization algorithm. However, because there can easily be thousands of weights and observations, most algorithms are too slow to be used in practice. This rules out second order methods like Newton's method which require a matrix of second derivatives. Even gradient descent (GD), which just calculates first derivatives and updates the weights at the $(t + 1)$th iteration via

$$
\begin{aligned}
w_{jk}^{t+1} &= w_{jk}^t - \eta \frac{\partial L}{\partial w_{jk}^t}, \\
v_{kl}^{t+1} &= v_{kl}^t - \eta \frac{\partial L}{\partial v_{kl}^t},
\end{aligned}
\tag{9.14}
$$

where η is the *learning rate*, can be too slow. The remainder of the section will discuss how to modify GD to speedup the training process.

We begin by tackling the large number of observations problem. Because the loss function is a sum of N observations, to make a single update to a single weight we must compute N derivatives. If $N = 1,000$ and there are $1,000$ weights, then we need to compute a million derivatives to update the weights just once. We then need to repeat this process enough times so we converge to a suitable local minimum.

A much faster method updates the weights by approximating the true gradient using smaller, but representative mini-batches of size $n < N$. That is, we approximate

$$\nabla L(\mathbf{W}, \mathbf{V}) = \frac{1}{N} \sum_{i=1}^{N} \nabla (\mathbf{y}_i - \widehat{\mathbf{y}}_i)^2 \approx \frac{1}{n} \sum_{i'=1}^{n} \nabla (\mathbf{y}_{i'} - \widehat{\mathbf{y}}_{i'})^2 \tag{9.15}$$

for a random selection of i'. This allows us to update the weights after computing only n derivatives. Next, we draw another mini-batch from the remaining observations, ensuring all observations in the dataset are used equally, and update the weights again. We repeat this for N/n batches, after which we have completed an *epoch* of training using all the data once.[5] We can then complete more epochs, each time randomizing the N/n batches, until suitable

---

[4] This sometimes referred to as the *cost function*.

[5] If N/n is not an integer, there are $\lceil N/n \rceil$ batches, the last batch being smaller than n.

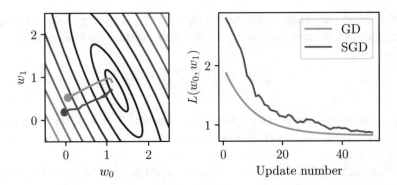

Fig. 9.14 Comparison between gradient descent and its stochastic variant. The left panel plots contours of the loss function and the updated weight paths, initially randomized near zero. The right panel shows the loss after each update

convergence is reached. Though we will not step down in the "best" direction each time, we will step much more frequently in a good directions, on average. This procedure is known as *stochastic gradient descent* (SGD). The mean squared error is useful for SGD because it decouples the learning rate η from the batch size. If we used sum of squares instead, larger batches would learn faster for a fixed η.

The difference between SGD and ordinary GD is illustrated in Fig. 9.14 for a simple network with just two weights so that $L = L(w_0, w_1)$. GD evolves perpendicularly to the contour lines while SGD meanders about, though they both approach similar small values of L after the same number of updates. Not pictured is the considerable speedup associated with SGD because multiple updates are made per epoch of training.

9.3.4   Learning the Game of Life Rules

We will now demonstrate how to build and use neural networks with the Keras library. In Sect. 9.1.1, we showed how complex global patterns can emerge in the Game of Life from a simple set of local rules. Supposing we did not know the rules, could we infer them based on grid snapshots, e.g. a series of snapshots like in Fig. 9.2? Although we know the rules end up being simple, it is not so obvious from just a series of snapshots. It is conceivable to have a modified Game of Life where a single cell can interact with cells within a larger radius. We will attempt to solve this problem using a neural network.

First we must define the inputs and output. Given a $n \times m$ grid of cells in the Game of Life, a neural network that has successfully learned the rules will be able to perfectly predict which cells live and die at the next timestep. The input data consists of an $n \times m$ array of binary elements, indicating whether the cell is dead or alive. For simplicity, we will update grids with constant boundary conditions, but only use the inner cells for training since the outer ones are externally forced to be dead, i.e. the rules are modified at the boundaries. The following generates the 1,000 samples of training data to input into Keras:

```
In [14]: from tensorflow import keras

          def add_grid_padding(grid):
              return np.pad(grid, pad_width=1, mode='constant', constant_values=0)

          def remove_grid_padding(grid):
              return grid[1:-1, 1:-1]

          def generate_data(n_sample, n_row, n_col):
              X, Y = [], []
              for _ in range(n_sample):
                  grid = add_grid_padding(
                      np.random.randint(low=0, high=2, size=(n_row, n_col)))
                  updated_grid = update_grid(grid)
                  X.append(remove_grid_padding(grid))
                  Y.append(remove_grid_padding(updated_grid))
              return np.expand_dims(X, -1), np.expand_dims(Y, -1)

          keras.utils.set_random_seed(1)
          n_row, n_col = 20, 20
          X_train, Y_train = generate_data(n_sample=1000, n_row=n_row, n_col=n_col)
          X_train.shape
```

Out[14]: (1000, 20, 20, 1)

Note that we have expanded each of the 1,000 grids to have shape (20, 20, 1) for Keras. The last dimension corresponds to the number of channels the input has. In our case, our grid can be thought of as a single grayscale "image" with one channel.

We are now ready to train a neural network using a fully connected architecture described by Fig. 9.12 in Sect. 9.3.2. We specify the 3-layer architecture with a `keras.Sequential()` model:

```
In [15]: input_shape = (n_row, n_col, 1)
          model = keras.Sequential([
              keras.Input(shape=input_shape),
              keras.layers.Dense(units=100, activation='relu'),
              keras.layers.Dense(units=1, activation='sigmoid')
          ])
```

We first specify the shape of the input data, add a hidden layer using ReLU activation with 100 neurons, and finally add an output layer using sigmoid activation for two distinct classes.[6]

Given an instantiated model, we can now configure training methods. We will use mean squared error as our loss function, optimize via stochastic gradient descent, and use accuracy as our performance metric. We will then iterate over the 1,000 samples in entirety in 10 epochs, each iteration using a batch size of 10.

```
In [16]: SGD = keras.optimizers.SGD(learning_rate=1)
          model.compile(loss='mse', optimizer=SGD, metrics=['accuracy'])
          model.fit(X_train, Y_train, epochs=10, batch_size=10, verbose=1)
```

---

[6] A binary classification only needs one unit since given $P(X = 0)$, we know $P(X = 1)$. In general, k-class classification requires $k - 1$ units.

```
Out[16]:  Epoch 1/10
          100/100 [====================] - 1s 6ms/step - loss: 0.2104 - accuracy: 0.6888
          Epoch 2/10
          100/100 [====================] - 1s 7ms/step - loss: 0.2084 - accuracy: 0.6896
          ...
          Epoch 10/10
          100/100 [====================] - 1s 7ms/step - loss: 0.2084 - accuracy: 0.6896
```

After 10 epochs of training, the network's prediction accuracy on the training data is better than random guessing, but is far from perfect and suggests it has not learned the rules of the Game of Life. We could try increasing the number of epochs, but the loss function the network is minimizing barely changes from the second to the tenth epoch. Perhaps the learning rate is too fast and the minima of interest are skipped over, perhaps the batch size is too small, or perhaps the loss function is ill-suited to measure a binary response. Before we make those, and countless other potential adjustments, it is illuminating to try to understand what the network has learned. In general, this extremely difficult due to the shear number of parameters estimated. In this example, however, it is useful to observe the predictions made by invoking the `predict()` method. Because we are using a sigmoid activation function, the output of the network is the probability a cell is alive. It turns out that predictions across the entire 1,000 training samples only result in two unique probabilities:

```
In [17]:  np.unique(model.predict(X_train))
```

```
Out[17]:  array([0.22743931, 0.36943513], dtype=float32)
```

With a symmetric thresholding rule of alive if $p > 0.5$ and dead otherwise, the model is predicts that every cell in the Game of Life will die at the next iteration. In other words, a model that simply predicts every cell dies has the same accuracy as the neural network. We can verify this by counting the number of dead cells in the next-iteration training data:

```
In [18]:  np.mean(Y_train == 0)
```

```
Out[18]:  0.6896
```

While underwhelming, the network has learned two features: it has learned that there are more dead cells than alive ones and, from the comparatively larger probability 0.37, that alive cells are more likely to stay alive. But it has clearly failed our goal to learn the rules of the Game of Life.

What went wrong? The trained network resulted in a solution that is just a wide local minima far from the truly informative ones. It turns out that the Game of Life is difficult, but not impossible, for neural networks to learn (see Springer and Kenyon 2021). One of the issues is our *fully connected* architecture. We know in the Game of Life, a cell's neighbors also determine its future state. Yet from the network output's two unique probabilities, we see it has only used information from a single cell to determine its future. If we used an architecture that clustered neighboring cells together, it would facilitate learning of larger patterns. A variation on the fully connected architecture, known as a *convolutional neural network* (CNN), does exactly this. We first demonstrate by example:

```
In [19]:  model = keras.Sequential([
              keras.Input(shape=input_shape),
              keras.layers.Conv2D(filters=3, kernel_size=(3, 3), activation='relu',
                                  padding='same'),
              keras.layers.Dense(units=1, activation='sigmoid')
          ])
          model.compile(loss='mse', optimizer=SGD, metrics=['accuracy'])
          model.fit(X_train, Y_train, epochs=10, batch_size=10, verbose=1)
```

```
Out[19]:  Epoch 1/10
          100/100 [====================] - 1s 2ms/step - loss: 0.1688 - accuracy: 0.7185
          Epoch 2/10
          100/100 [====================] - 0s 2ms/step - loss: 0.1421 - accuracy: 0.7248
          ...
          Epoch 10/10
          100/100 [====================] - 0s 2ms/step - loss: 0.0028 - accuracy: 1.0000
```

This neural network has successfully learned the rules of the Game of Life with just one change in the second layer, using `Conv2D` instead of `Dense` (fully connected). We should verify that the network generalizes to external data outside the training set:

```
In [20]:  X_test, Y_test = generate_data(n_sample=1000, n_row=n_row, n_col=n_col)
          print(model.metrics_names)
          print(model.evaluate(X_test, Y_test, verbose=0))
```

```
Out[20]:  ['loss', 'accuracy']
          [0.002449536696076393, 1.0]
```

Indeed, its perfect prediction accuracy is maintained.

The `filters` argument sets the flexibility of the CNN and is analogous to the number of neurons in a fully connected hidden layer. The `kernel_size` specifies the clustering size. In theory, a 3×3 clustering size would be ideal since the Game of Life rules are based on 3×3 neighborhoods, but in practice we do not know the ideal size ahead of time (assuming it exists). Despite the 5×5 kernel size, the CNN able to properly filter and learn desired rules.[7]

The Game of Life has shown us once again that despite having a simple set of rules, the evolution of the game is complex enough such that the rules are hard for a vanilla neural network to deduce. That being said, we have seen that neural networks can represent complex nonlinear functions and that we can modify them so that they are able to learn these complex patterns. A myriad of other neural network extensions exist that address shortcomings we have not seen and speak to the potential of these methods. However, questions like choosing the correct architecture, tuning the parameters, and interpreting the trained network are not so simple to answer. Neural networks are certainly a powerful tool, but are not a panacea for all problems and do not replace intuition and value built from simpler theoretical models.

---

[7] This property makes CNNs particularly common in image recognition. Instead of mapping a single pixel to the output classification, neighboring pixels are grouped together which can reduce noise and overfitting.

9.4   Exercises

9.1. Exploring the Game of Life.

(a) Play the Game of Life starting from 4-cell configurations in the shape of Tetris blocks. Ensure there is enough padding so there are no edge effects. Characterize the results as in Fig. 9.1.

(b) Consider a grid with just 2 gliders moving in opposite directions. Define a collision as an event that changes the gliders as they pass each other. Approximately how many unique collisions are there? What is the most common outcome of a collision?

(c) Determine the time evolution of the initial configuration in Fig. 9.15 on an infinite grid. Assume all other cells start dead. Does the system ever stabilize?

9.2. Game of Life's sensitivity to initial conditions. Start with the grid in Fig. 9.2 at $t = 0$, generated by

```
In [21]:   np.random.seed(1)
           grid = np.pad(np.random.randint(low=0, high=2, size=(20, 20)), 1)
```

It eventually stabilizes at $t = 105$. Is the same stable configuration reached if 1 cell at $t = 0$ was flipped? Consider all 20×20 unique initial conditions that differ from grid by just 1 cell. How many of these initial conditions result in the same stable configuration at $t = 105$? More generally, for each initial condition, calculate the number of cells at $t = 105$ that differ from grid's eventual $t = 105$ stable configuration and plot the distribution of differences. What distinguishes initial conditions that end up with large differences as opposed to small but nonzero differences?

9.3. Game of Life with periodic boundary conditions. Modify the text's Game of Life implementation to use periodic boundary conditions. Specifically, the grid is wrapped so any cell at the right boundary has neighbors on the left boundary and any cell at the

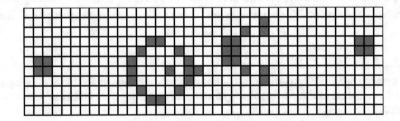

Fig. 9.15 Initial configuration for Exercise 9.1c

top boundary has neighbors on the bottom boundary. It follows that all corner cells are connected. Test your code by visualizing the behavior of a single glider.

Using the same initial configuration of Fig. 9.2 (without the dead cell padding), compare the final stabilized grids.

*9.4. Macroscopic properties and long term behavior of random grids. A random grid can be formalized as a grid whose individual cells start alive with probability p, independent of other cells. The random grid in Fig. 9.2 set $p = 0.5$. In this problem, you will investigate properties of random grids as a function of p using periodic boundary conditions (Exercise 9.3).

(a) Write a function that initializes a random grid with arbitrary start probability $p \in [0, 1]$.

(b) For a given grid, write a function to determine the number of timesteps until it stabilizes, i.e. dies out or is composed of still lifes, oscillators, and/or spaceships whose behaviors predictably repeat. Since some grids may not stabilize for a long time, impose a maximum timesteps parameter T_{max}. Calibrate this parameter for a random 20×20 grid. Does it depend strongly on p? Restrict analysis in future parts to random 20×20 grids that stabilize within T_{max}.

(c) Using Monte Carlo simulations, plot the proportion of grids that die out, stabilize into a collection of still lifes, and stabilize with at least one oscillator or spaceship as a function of p. Run enough simulations so that standard errors are within 2%.

(d) Plot the average ratio of alive to total cells when a grid first stabilizes as a function of p. These are estimates of the asymptotic densities of cells. Run enough simulations so that standard errors are within 0.1%.

(e) Do grids that stabilize quickly differ from those that take a long time to stabilize? What about when p is fixed, say $p = \frac{1}{2}$?

(f) Investigate temporal behavior of cell densities. Plot the average density over time for various p with distinct asymptotic behavior. Describe your results. Are they consistent with what you expected?

9.5. Game of Life on a hexagonal grid. There are many variations of the Game of Life obtained by changing the rules and/or geometry. Here you will implement 2 different rulesets on a hexagonal grid.

(a) One way to represent a hexagonal grid uses a 2D array where every 2nd row is offset. A 3×3 array corresponds to the hexagonal grid in Fig. 9.16 where each hexagonal cell is labeled by the corresponding array indices. Modify `update_grid()` accordingly with the same cell update rules except now each cell has 6 instead of 8 neighbors.

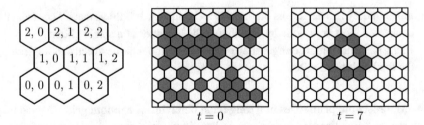

Fig. 9.16 Representation of a hexagonal grid using a 2D array and test configuration for Exercise 9.5

(b) Modify `play()` to visualize hexagonal grids like in Fig. 9.16. There is no `pcolormesh()` equivalent for hexagonal grids, so one approach plots each hexagon individually as a patch. A list of patches constitutes a frame and can then be appended to `ims`. For instance, to plot a single hexagonal patch at (x, y) with radius r on an `Axes ax`, use

```
In [24]:  from matplotlib.patches import RegularPolygon
          hexagon = RegularPolygon((x, y), numVertices=6, radius=r)
          ax.add_patch(hexagon)
```

Test your code by playing the $t = 0$ configuration in Fig. 9.16. It should converge to the still life at $t = 7$. The grid can be initialized with

```
In [25]:  np.random.seed(1)
          grid = np.pad(np.random.randint(low=0, high=2, size=(8, 8)), 1)
```

(c) Characterize a few common still lifes. What is the largest still life you come across or can think of? Can you find an oscillator?

(d) Modify the update rules to your liking. For instance, one might object to the original Game of Life update since there are now 2 fewer neighbors. Characterize common patterns that arise.

9.6. Variation on parameters in the Nagel–Schreckenberg model.

(a) Consider the deterministic limit $p_{slow} = 0$ with length 100 and $v_{max} = 5$. Plot the time averaged flow as a function of density. What kind of transition occurs and at what critical density ρ_*? For $\rho > \rho_*$, after initial transients, why does each car in a spacetime plot appear to translate 1 cell backward per timestep? What is different for $\rho < \rho_*$?

(b) Modify the speed limit to two extremes: $v_{max} = 1$ and $v_{max} = 25$. How does the interpretation of the model change? What differences, if any, do you observe in spacetime and average flow vs. density plots?

9.7. Nagel–Schreckenberg model with open boundaries. Label a length L road's cells by $x = 0, 1, \ldots, L - 1$. Open boundary conditions are more complex because there is no well defined interaction for the car closest to $x = L - 1$. For example, there may be a car in front at $x = L$ we cannot see that should slow the car. Furthermore, as cars exit the road, the number and thus density of cars we observe changes. Adding a car at $x = 0$ whenever a car exits is problematic because there may already be a car there. Introducing probabilistic entrance and exit rates is one remedy you will explore here.

(a) Modify the text's Road.update() method to handle open boundaries as follows. Before making any updates, if there is not a car at $x = 0$ add one with $v = v_{max}$ with probability p_{ent}. Update all cars except for the one closest to $x = L$ as usual. For the boundary car, assume there is no car in its path. Then if its updated position $x' \geq L$, set $x' = L - 1$ with velocity $v' = 0$ and remove it with probability p_{exit}.

(b) Make an analogous pair of spacetime plots to those in Fig. 9.4 with $v_{max} = 5$ and $p_{slow} = 0.1$, choosing parameters p_{ent} and p_{exit} so that densities are comparable. Comment on your results.

(c) Since density is no longer a fixed property of the road, a useful alternative measure is the time averaged density—the average number of cars that occupy a fixed cell over T timesteps. This definition is similar to the time averaged flow. Plot the average flow as a function of average density. Identify the critical (time averaged) density $\overline{\rho}_*$ transitioning from free flow to traffic jams. Does $\overline{\rho}_*$ depend on which cell you measure?

9.8. Investigating period doubling. We have seen that a cascade of period doublings in the logistic map leads to chaos. In this exercise, you will find where period doublings occur and show they lead to chaos in other periodic windows.

(a) Code an algorithm that determines the period of a sequence of values generated by the logistic map $[x_0, x_1, \ldots, x_T]$ if there is a periodic orbit. Keep in mind that some orbits may converge very slowly, so build a reasonably small precision level into your algorithm.

(b) Test your algorithm just before $r = \frac{3}{4}$ where the first period doubling occurs, say $r = 0.749$. Does your algorithm correctly detect a period of 1? If not, why, and how can you fix it? *Hint:* A plot of x_t over time may be helpful. What is the difference between values separated by 1 and 2 timesteps and how do they compare to your precision level?

(c) Modify your algorithm to accurately determine where a period doubling point r_p occurs, where p is the associated period of the orbits after doubling. What parameters control how precise of an r_p value you can obtain? Obtaining $r_2 = \frac{3}{4}$ computationally is a good benchmark. *Hint:* A simple, but efficient implementation can be based off the bisection algorithm.

(d) Obtain $r_2, r_4, r_8, \ldots, r_{64}$ to at least 4 digit precision. Then compute the ratio of distances between successive r_p points to estimate Feigenbaum's constant, e.g. $(r_{32} - r_{16})/(r_{64} - r_{32})$. How close do you get to $\delta = 4.669...$? Using your estimates, predict where r_{256} is and try to find it with your algorithm. Estimate where r_∞ begins, i.e. chaos.

(e) Repeat **(d)** for the period 3 window beginning near $r = 0.96$, finding $r_3, r_6, \ldots, r_{96}$ and an estimate of δ. Using your data, predict where the window ends, and compare with an estimate made from just a bifurcation diagram.

9.9. Self-similarity in the logistic map.

(a) Modify `bifurcation_plot()` so you can generate a bifurcation diagram on any magnified region $[x_{min}, x_{max}] \times [r_{min}, r_{max}]$. Ensure in the magnified region there are n_keep points plotted. Use your modification to magnify any branch of the period 3 window near $r = 0.96$, the largest of the periodic windows after the onset of chaos. Add another level of magnification in the resultant largest window. Compare your plots to Fig. 9.8 (right).

(b) Self-similarity is not unique to the period 3 window. There is a period 11 window beginning near $r = 0.92043$. Find it, then magnify any of the 11 branches.

9.10. Other chaotic maps in 1D. For each map, compute equilibrium points of period 1 and characterize their stabilities, create an interactive evolution widget, and plot the bifurcation diagram. Compare and contrast your results with the logistic map.

(a) Explore the sine map:

$$x_{t+1} = r \sin(\pi x_t), \quad r \in [0, 1], \; x_t \in [0, 1].$$

Viewed as a Taylor expansion about $\pi/2$, the sine map contains nonlinearities at all even powers, not just quadratic.

(b) Explore the tent map:

$$x_{t+1} = r \times \begin{cases} x_t/s & \text{if } x_t < s, \\ (1 - x_t)/s & \text{if } x_t > 1 - s, \\ 1 & \text{otherwise,} \end{cases}$$

a two parameter map, $r \in [0, 1]$ and $s \in (0, \frac{1}{2}]$, defined for $x_t \in [0, 1]$. Its name comes from its shape at $s = \frac{1}{2}$. What does the parameter s do? Consider the $s = \frac{1}{2}$

case and $s = 0.49$ case separately. Why are there no periodic windows after chaos emerges when $s = \frac{1}{2}$?

9.11. The XOR problem. Consider the simple neural network in Fig. 9.10. We demonstrated that with binary inputs x_1 and x_2 and proper choice of parameters w_1, w_2, and b, the network effectively acts as a logical and operator, i.e. $y = 1$ if $x_1 = x_2 = 1$ and $y = 0$ otherwise. Show that this simple network architecture cannot replicate the behavior of an xor operator (exclusive or), i.e.

$$y = \begin{cases} 1 & \text{if } (x_1, x_2) \in \{(1,1), (0,0)\}, \\ 0 & \text{if } (x_1, x_2) \in \{(0,1), (1,0)\}. \end{cases}$$

Introduce a hidden layer with ReLU activation and specify an architecture with sample parameter values that replicates an xor operator.

9.12. Neural networks as function approximators.

(a) Generate n samples of training data according to

$$x \sim \mathcal{N}(0, 1),$$
$$y = x + \mathcal{N}(0, 1),$$

where $\mathcal{N}(\mu, \sigma^2)$ represents a random value from a normal distribution with mean μ and variance σ^2, e.g. x = np.random.randn(n). We are essentially generating data that fits the line $y = x$, but with random fluctuations. Train a fully connected neural network with your chosen n. Plot the original data and overlay two curves: the theoretical solution $y = x$ and the predictions of your trained network. Now repeat for various n. What is the (approximate) minimum sample size n such that the neural network generally represents the true line?

(b) Repeat **(a)**, but for a more complex Gaussian:

$$x \sim \mathcal{N}(0, 1),$$
$$y = e^{-x^2} + \mathcal{N}(0, 1),$$

(c) Now extend to two dimensions, i.e. x = np.random.randn(n, n), and vary between two complexities: a plane and a 2D Gaussian. Approximately how much larger does the minimum sample size need to be?

*(d) Extend further to ten dimensions. Since we cannot visualize how well a 10D Gaussian fits, you will need to determine a metric to measure the neural network's performance. How does the (approximate) minimum sample size scale as the dimensionality increases?

9.13. Image recognition. Visual tasks like recognizing handwritten digits are simple for humans, but very difficult for logic based programs. Neural networks circumvent the need to create a large set of rules, with many exceptions to accommodate different styles, though there is one important caveat—they require lots of quality training data.

(a) We start with a classic example using the MNIST dataset of handwritten digit images. Load in the data with the following:

```
In [27]:  (X_train, y_train), (X_test, y_test) = (
              keras.datasets.mnist.load_data())
          Y_train = keras.utils.to_categorical(y_train)
          Y_test = keras.utils.to_categorical(y_test)
```

Get familiar with the data and visualize a few images. What does the `to_categorical()` utility do, and why do we need it before training a neural network?

(b) Train a feed forward neural network with a single hidden layer to recognize digits from an image. Evaluate its accuracy on the test data. Once you have a network that does better than random guessing, spend some time optimizing the network for accuracy. Suggestions include tuning the parameters, changing the architecture, or trying a convolutional neural network. Explain why your changes had the effect they did. Once you have a network that works well, visualize a few images where its predictions are incorrect, and comment on your findings.

(c) Let us revisit our Game of Life example from a different angle. Instead of predicting the next state of the game, suppose we want to answer a more general question: given an image of the current state of the game, classify whether the board will stabilize within 100 timesteps. Do you expect this to be more or less difficult than learning the rules of the game? Justify your answer.

*(d) Implement the classifier described in (c). See Exercise 9.4 for more details on generating the data.

References

K. Nagel and M. Schreckenberg. A cellular automaton model for freeway traffic. *Journal de Physique I*, 2(12):2221–2229, 1992.

J. M. Springer and G. T. Kenyon. It is hard for neural networks to learn the game of life. In *2021 International Joint Conference on Neural Networks (IJCNN)*, pages 1–8. IEEE, 2021.

S. H. Strogatz. *Nonlinear dynamics and chaos.* (Westview Press, Cambridge, MA), 1994.

J. Wang. *Computational modeling and visualization of physical systems with Python.* (Wiley, New York), 2016.

Program Listings

A

This appendix lists most complete Python (`.py`) programs discussed in the text. All programs, including in Jupyter notebook (`.ipynb`) format, are available online at

- https://github.com/com-py/intro

  Additionally, select GlowScript and Trinket programs are available at

- GlowScript: https://www.glowscript.org/#/user/jaywang/folder/intro/
- Trinket: https://github.com/com-py/intro

A.1   Programs from Chap. 3

Program A.1.1 Interactive computing with IPywidgets.

```
1  # See links to online programs below
```

See Chap. 3 Program `p1-ipywidgets` online at https://github.com/com-py/intro.

Program A.1.2 Interactive computing with Matplotlib widgets.

```
1  import matplotlib.pyplot as plt
   from matplotlib.widgets import Slider, RadioButtons, CheckButtons
3  import numpy as np

5  x = np.linspace(-1, 1, 101)      # grid

7  def updatefig(*args):            # update figure data
       k = slider.val
9      if func.value_selected == fs[0]:        # sin
           f = np.sin(k*np.pi*x)
```

© The Editor(s) (if applicable) and The Author(s), under exclusive license to Springer   233
Nature Switzerland AG 2023
J. Wang and A. Wang, *Introduction to Computation in Physical Sciences*,
Synthesis Lectures on Computation and Analytics,
https://doi.org/10.1007/978-3-031-17646-3

```
11       else:
             f = np.cos(k*np.pi*x)
13       if refresh.get_status()[0]:        # checkbox 0 val
             plot.set_ydata(f)
15           ax.set_ylim(min(f), max(f))
     fig.canvas.draw_idle()
17
   fig = plt.figure()
19 fig.canvas.set_window_title('Interactive plotting')
   fig.subplots_adjust(left=0.15, bottom=0.35, top=.95)
21 ax = fig.add_subplot(111)
   plot = ax.plot(x, [0]*len(x))[0]
23
   axes = fig.add_axes([.15, .20, .75, .03])        # k slider
25 slider = Slider(axes, '$k$', 0, 2, 1)             # range, init val
   slider.on_changed(updatefig)
27
   fs  = ['sin', 'cos']         # radio buttons
29 rbox =fig.add_axes([.15, .02, .15, .15])
   func = RadioButtons(rbox, fs, 0)     # active = sin (0)
31
   cbox = fig.add_axes([.4, .02, .15, .15])            # checkbox, real+extra
33 refresh = CheckButtons(cbox, ['refresh', 'extra'], [1,0])
   func.on_clicked(updatefig)
35 refresh.on_clicked(updatefig)
37 updatefig(None)
   plt.show()
```

Program A.1.3 Animation with Matplotlib.

```
# See links to online programs below
```

See Chap. 3 Program p3-matplotam.ipynb online at https://github.com/com-py/intro.

Program A.1.4 3D visualization with VPython. Also available in GlowScript.

```
1 import vpython as vp
3 scene = vp.canvas(title=('right drag to change camera,' +
                           'double drag to zoom'))
5 ball = vp.sphere()
  box = vp.box(pos=vp.vector(0,0,-1.1), length=10, height=5, width=.2,
7                color=vp.color.cyan)
9 t = 0
  while t<10:
11     vp.rate(10)
       ball.pos = vp.vector(2*vp.sin(2*t),0,0)
13     t = t + 0.1
```

Program A.1.5 Symbolic computation with Sympy.

```
1 # See links to online programs below
```

See Chap. 3 Program p5-sympy.ipynb online at https://github.com/com-py/intro.

Program A.1.6 Numpy library.

```
1 # See links to online programs below
```

See Chap. 3 Program `p6-numpy.ipynb` online at https://github.com/com-py/intro.

Program A.1.7 Scientific computing with Scipy.

```
1 # See links to online programs below
```

See Chap. 3 Program `p7-scipy.ipynb` online at https://github.com/com-py/intro.

Program A.1.8 Numba optimization.

```
1  import numpy as np, time
   from numba import jit
3
   @jit                          # optimization
5  def calc(u0, u1):
       u2 = - u0
7      for n in range(1, N-1):
           u2[n] += u1[n-1] + u1[n+1]        # left, right
9      return u2

11 N = 5000
   u0 = np.random.random(N)
13 u1 = 0.9*u0

15 tc = time.perf_counter()
   for i in range(2000):
17     u2 = calc(u0, u1)
       u0, u1 = u1, u2
19
   tc = time.perf_counter() - tc
21 print(tc)
```

A.2   Programs from Chap. 4

Program A.2.1 Curve fitting.

```
1  from scipy.optimize import curve_fit
   import matplotlib.pyplot as plt
3  from numpy.random import random
   import numpy as np
5
   def lin(t, a, b):
7      return a + b*t

9  def quad(t, a, b, c):
       return a + b*t + c*t*t
11
   N = 101
13 t = np.linspace(0, 0.5, N)
   x = 5*t - 4.9*t*t + 0.2*random(N)
```

```
15
   cl = curve_fit(lin, t, x)[0]
17 cq = curve_fit(quad, t, x)[0]
   print(cl, cq)
19
   plt.plot(t, x, 'o')
21 plt.plot(t, lin(t, cl[0], cl[1]), 'k--')
   plt.plot(t, quad(t, cq[0], cq[1], cq[2]), 'r-')
23 plt.xlabel('t (s)'), plt.ylabel('f (m)')
   plt.show()
```

Program A.2.2 Locating minima and maxima.

```
   import numpy as np
2  import matplotlib.pyplot as plt

4  def minmax(y):          # return indices of minima and maxima
       mini = (y[1:-1]<=y[:-2])*(y[1:-1]<=y[2:])    # minima array
6      maxi = (y[1:-1]>=y[:-2])*(y[1:-1]>=y[2:])    # maxima array
       mini = 1 + np.where(mini)[0]                 # indices of minima
8      maxi = 1 + np.where(maxi)[0]                 # indices of maxima
       return mini, maxi
10
   x = np.linspace(0, 5, 100)
12 y = np.exp(-x/2)*np.sin(2*np.pi*x)
   mini, maxi = minmax(y)
14
   plt.plot(x, y, '-o', mfc='none')                # open circle marker
16 plt.plot(x[mini], y[mini], 'kv', label='minima')
   plt.plot(x[maxi], y[maxi], 'r^', label='maxima')
18 plt.legend()
   plt.xlabel('x')
20 plt.ylabel('y')
   plt.show()
```

A.3   Programs from Chap. 5

Program A.3.1 Ideal projectile motion.

```
1  import matplotlib.pyplot as plt
   import numpy as np
3  g = 9.8
   a = np.array([0.0, -g])
5  r = np.array([0.0, 0.0])
   v = np.array([4.0, 9.0])
7  dt = 0.1
   t = 0.0
9  tlist = []
   rlist = []
11 vlist = []
   for i in range(20):
```

```
13      tlist.append(t)
        rlist.append(r)
15      vlist.append(v)
        r = r + v*dt
17      v = v + a*dt
        t = t + dt
19
   rlist = np.array(rlist)
21 vlist = np.array(vlist)

23 plt.plot(tlist, rlist[:,0], tlist, rlist[:,1])
   plt.xlabel('time (s)')
25 plt.ylabel('x, y (m)')
   plt.figure()
27 plt.plot(tlist, vlist[:,0], tlist, vlist[:,1])
   plt.xlabel('time (s)')
29 plt.ylabel('vx, vy (m/s)')
   plt.show()
```

Program A.3.2 Realistic projectile motion with drag. Also available in GlowScript.

```
   import matplotlib.pyplot as plt
2  import numpy as np
   c1 = 0.0
4  c2 = 0.2
   g = 9.8
6  ag = np.array([0.0, -g])
   r = np.array([0.0, 0.0])
8  v = np.array([4.0, 9.0])
   dt = 0.1
10 t = 0.0
   tlist = []
12 rlist = []
   vlist = []
14 while r[1] >= 0.0:          # y-position above 0
        tlist.append(t)
16      rlist.append(r)
        vlist.append(v)
18      speed = np.sqrt(v[0]**2 + v[1]**2)
        a = ag - c1*v - c2*speed*v
20      r = r + v*dt
        v = v + a*dt
22      t = t + dt

24 rlist = np.array(rlist)
   vlist = np.array(vlist)
26
   plt.plot(tlist, rlist[:,0], tlist, rlist[:,1])
28 plt.xlabel('time (s)')
   plt.ylabel('x, y (m)')
30 plt.figure()
   plt.plot(tlist, vlist[:,0], tlist, vlist[:,1])
32 plt.xlabel('time (s)')
   plt.ylabel('vx, vy (m/s)')
34 plt.show()
```

Program A.3.3 Simple harmonic oscillator in 2D. Also available in GlowScript.

```
   import matplotlib.pyplot as plt
 2 import numpy as np
   omega = 1.0
 4 r = np.array([1.0, 1.5])
   v = np.array([1.0, 0.0])
 6 dt = 0.1
   t = 0.0
 8 tlist = []
   rlist = []
10 while t< 10.0:
       tlist.append(t)
12     rlist.append(r)
       r = r + v*dt
14     a = - omega**2 * r
       v = v + a*dt
16     t = t + dt

18 rlist = np.array(rlist)

20 plt.plot(tlist, rlist[:,0], tlist, rlist[:,1])
   plt.xlabel('time (s)')
22 plt.ylabel('x, y (m)')
   plt.figure()
24 plt.plot(rlist[:,0], rlist[:,1])
   plt.xlabel('x (m)')
26 plt.ylabel('y (m)')
   plt.show()
```

Program A.3.4 Planetary motion. Also available in GlowScript.

```
 1 import matplotlib.pyplot as plt
   import numpy as np
 3 import vpython as vp

 5 GM = 4*np.pi**2
   r = np.array([1.0, 0.0])
 7 v = np.array([0.0, 2*np.pi])
   dt = 0.01
 9 rlist = []
   t = 0.0
11 scene = vp.canvas(background=vp.vector(.0, .20, 1))
   star = vp.sphere(radius=0.1, color=vp.color.yellow)
13 planet = vp.sphere(radius=0.1, texture=vp.textures.earth)
   trail = vp.curve(radius=0.005)
15
   while t< 2.0:
17     vp.rate(20)
       planet.pos=vp.vector(r[0], r[1], 0.)
19     trail.append(planet.pos)
       rlist.append(r)
21     r = r + v*dt
       rmag = np.sqrt(r[0]**2 + r[1]**2)
23     a = - GM * r/rmag**3
       v = v + a*dt
25     t = t + dt
```

```
27  rlist = np.array(rlist)
    ax = plt.subplot(111, aspect='equal')     # equal aspect ratio
29  ax.plot(rlist[:,0], rlist[:,1])
    plt.xlabel('x (AU)')
31  plt.ylabel('y (AU)')
    plt.show()
```

Program A.3.5 Motion in electromagnetic fields.

```
    import matplotlib.pyplot as plt
2   from mpl_toolkits.mplot3d import Axes3D
    import numpy as np
4
    def paraperp(v, B):      # project para // and perp _|_ velocities
6       vpara = np.dot(v,B) * B/np.dot(B, B)
        vperp = v - vpara
8       mperp = np.sqrt(np.dot(vperp, vperp))
        return vpara, vperp, mperp
10
    E = np.array([0, 0.0, 1.e4])
12  B = np.array([1.0, 0.0, 0.0])
    qm = 1.0          # q/m
14  dt = 0.01
    t = 0.0
16  r = np.array([0., 0., 0.])
    v = np.array([2.e4, 1.e5, 0.])
18  rlist = []
    while t < 20.0:
20      rlist.append(r)
        r = r + v*dt/2          # half step r at midpoint
22
        v = v + qm*E*dt/2       # half step by E field
24      vpara, vperp, mperp0 = paraperp(v, B)
26      a = qm*np.cross(v, B)   # full step by B field
        v = v + a*dt
28      vpara, vperp, mperp = paraperp(v, B)
        vperp = vperp*mperp0/mperp      # correct magnitude
30      v = vpara + vperp
32      v = v + qm*E*dt/2       # 2nd half step by E field
        r = r + v*dt/2         # 2nd half step r
34      t = t + dt
36  rlist = np.array(rlist)/np.max(np.abs(rlist))    # scale r
    ax = plt.subplot(111, projection='3d')
38  ax.plot(rlist[:,0], rlist[:,1], rlist[:,2])
    ax.set_xlabel('x (arb)'), ax.set_ylabel('y (arb)')
40  ax.set_zlabel('z (arb)'), ax.set_zlim(-.5 ,0)
    plt.show()
```

A.4   Programs from Chap. 6

Program A.4.1 Symbolic determination of normal modes.

```
1  # See links to online programs below
```

See Chap. 6 Program p1-normalmodes.ipynb online at https://github.com/com-py/intro.

Program A.4.2 Coupled three-body oscillator.

```
1  import matplotlib.pyplot as plt
   import numpy as np
3
   k1, k2 = 1.0, 1.0
5  m1, m2, m3 = 1.0, 3.0, 1.0
   u = np.array([1.0, 0.0, 0.5])
7  v = np.array([0.0, 0.0, 0.0])

9  t, dt = 0.0, 0.1
   tlist, ulist = [], []
11 while t< 20.0:
       tlist.append(t)
13     ulist.append(u)

15     u = u + 0.5*v*dt              # leapfrog
       f12 = -k1*(u[0]-u[1])
17     f23 = -k2*(u[1]-u[2])
       a = np.array([f12/m1, -f12/m2 + f23/m2, -f23/m3])
19     v = v + a*dt
       u = u + 0.5*v*dt
21     t = t + dt

23 ulist = np.array(ulist)
   plt.plot(tlist, ulist[:,0], tlist, ulist[:,1], '-.', tlist, ulist[:,2], '--')
25 plt.xlabel('time (s)')
   plt.ylabel('$u_{1,2,3}$ (m)')
27 plt.show()
```

Program A.4.3 Waves on a string.

```
1  import matplotlib.pyplot as plt, numpy as np
   import matplotlib.animation as am
3  from numba import jit

5  @jit                          # optimization
   def waves(u0, u1):
7      u2 = 2*(1-b)*u1 - u0                    # unshifted terms
       for i in range(1, len(u0)-1):
9          u2[i] += b*( u1[i-1] + u1[i+1] )    # left, right
       return u2
11
   def updatefig(*args):                       # args[0] = frame
13     global u0, u1
       u2 = waves(u0, u1)
15     u0, u1 = u1, u2
```

```
          plot.set_data(x, u0)                     # update data
17
    N, b = 100, 1.0              # number of points, beta^2
19  x = np.linspace(0., 1., N)
    u0 = np.sin(np.pi*x)**2      # initial cond
21  u1 = 0.98*u0

23  fig = plt.figure()
    plot = plt.plot(x, u0)[0]                # create plot object
25  ani = am.FuncAnimation(fig, updatefig, interval=5)      # animate
    plt.axis('off')        # don't need axes
27  plt.ylim(-1.2,1.2)
    plt.show()
```

Program A.4.4 Standing waves on a string.

```
    from scipy.linalg import eigh        # eigenvalue solver
2   import numpy as np, matplotlib.pyplot as plt

4   N = 101            # number of grids
    x = np.linspace(0., 1., N)    # grids
6
    A = np.diag([2.]*(N-2))                                 # diagonal
8   A += np.diag([-1.]*(N-3),1) + np.diag([-1.]*(N-3),-1)   # off diags
    A = A/(x[1]-x[0])**2          # delta x = x_1 - x_0
10
    lamb, X = eigh(A)             # solve for eigenvals and eigenvecs
12
    sty = ['-','--','-.',':']
14  print (np.sqrt(lamb[:len(sty)])/np.pi)                  # wave vector k/pi

16  for i in range(len(sty)):    # plot a few standing waves
        Xi = np.insert(X[:,i], [0, N-2], [0., 0.])   # insert end points
18      plt.plot(x, Xi, sty[i], label=repr(i+1))

20  plt.xlabel('$x$'), plt.ylabel('$X$')
    plt.legend(loc='lower right'), plt.show()
```

A.5   Programs from Chap. 7

Program A.5.1 Time dilation.

```
1   import numpy as np, matplotlib.pyplot as plt
    import matplotlib.animation as am
3   from matplotlib.widgets import Slider

5   def updatefig(*args):                      # args[0] = time counter
        n = args[0]
7       if n < len(x):
            xn = x[n-1]
9           w = 0.1/gamma                              # contracted width
            boxx = [xn - w, xn - w, xn + w, xn + w]      # car
```

```
11          boxy = [0, .1, .1, 0]
            plot[0].set_data(x[:n], y[:n])                    # lab
13          plot[1].set_data([xn, xn], [y[n-10],y[n-1]])      # light beam
            plot[2].set_data(boxx, boxy)
15          labtxt.set_position((.9*L, .8))                   # time update
            labtxt.set_text(r'$\beta$='+repr(slider.val)[:4]     # 4 digits
17                        + '\n' + repr(tlab[n])[:4])
            cartxt.set_position((xn-0.2*w, .12))
19          cartxt.set_text(repr(tcar[n])[:4])

21 def animate(dummy):
       global x, y, tlab, tcar, L, gamma, labtxt, ani
23
       beta = slider.val
25     gamma = 1./np.sqrt((1-beta)*(1+beta))
       y = 1 - np.abs(np.linspace(-1,1,201))     # up/down parts
27     L = beta*gamma                            # half way point
       x = np.linspace(0, 2*L, len(y))           # grid
29     tcar = np.linspace(0, 2, len(y))          # proper time (train)
       tlab = gamma*tcar                         # time dilation (lab)
31     ax.set_xlim(-.05, max(.2,2.1*L)), ax.set_ylim(0,1)
       labtxt = ax.text(.9*L,.8, '', color='blue')   # lab time, new each run
33     ani = am.FuncAnimation(fig, updatefig, interval=1)

35 fig = plt.figure()
   fig.canvas.set_window_title('Time dilation')
37 fig.subplots_adjust(left=0.15, bottom=0.35, top=.95)    # setup for slider
   ax, z = fig.add_subplot(111), [0]
39 plot = ax.plot(z, z, '-', z, z, '-or', z, z, '-r')      # 3 curves
   ax.set_xlabel('$x$ (arb)'), ax.set_ylabel('$y$ (arb)')
41 cartxt = ax.text(.2, .2, '', color='red')               # train time

43 axes = fig.add_axes([.15, .20, .75, .03])               # beta slider
   slider = Slider(axes, r'$\beta$', 0, .99, .5)            # range, init val
45 slider.on_changed(animate)        # link slider to animate
   animate(None)
47 plt.show()
```

Program A.5.2 Double-slit interference.

```
1 import numpy as np, matplotlib.pyplot as plt
  from matplotlib.widgets import Slider, RadioButtons
3
  def singleslit(lamda, x0, w, h, N=10):
5     k = 2*np.pi*1.e9/lamda        # wave vector (conv from nm)
      theta = np.arctan(x/h)
7     r = np.sqrt(h*h + x*x)
      d = np.linspace(-w/2, w/2, N)   # divisions
9     psi = 0.0
      for s in d - x0:     # offset origin
11        r1 = np.sqrt(r*r +s*s - 2*r*s*np.sin(theta))
          psi += np.exp(1j*k*r1)/r1
13    return psi/N          # normalize

15 def updatefig(dummy):        # called when paras are changed
       lamda, w, d = sliders[0].val, sliders[1].val, sliders[2].val
17     psi = singleslit(lamda, -1.e-6*d/2, 1.e-6*w, h)    # slit 1
```

```
19      if button.value_selected == items[1]:     # slit 2
            psi = (psi + singleslit(lamda,  1.e-6*d/2, 1.e-6*w, h))/2
        intensity = np.abs(psi)**2
21      plot.set_ydata(intensity)
        ax.set_ylim(0, max(intensity))
23      fig.canvas.draw_idle()

25  h = 1.0      # dist to screen (m)
    x = np.linspace(-.1*h, .1*h, 1000)     # grids on screen
27  vars = [r'$\lambda$, nm', 'w, $\mu$m', 'd, $\mu$m']     # lamda (nm), w, d (micron)
    vals = [[400, 800, 600], [0, 20, 10], [0, 100, 50]]     # min, max, start
29
    fig = plt.figure()
31  fig.canvas.set_window_title('Double slit interference')
    ax = fig.add_subplot(111)
33  fig.subplots_adjust(left=0.15, bottom=0.35, top=.95)
    plot = ax.plot(x, 0*x, '-r', lw=2)[0]     # initiate plot
35  plt.xlabel('$x$ (m)'), plt.ylabel('Intensity (arb.)')

37  sliders = []                               # draw sliders at bottom
    xw, yw, offset, dx, dy = .15, .20, .05, .5, .03     # widgets pos.
39  for i in range(len(vars)):
        axes = fig.add_axes([xw, yw-i*offset, dx, dy])
41      sliders.append(Slider(axes, vars[i], vals[i][0], vals[i][1], vals[i][2]))
        sliders[i].on_changed(updatefig)
43
    items = ['single slit', 'double slit']   # add buttons
45  baxes =fig.add_axes([.75, .1, .15, .15])
    button = RadioButtons(baxes, items, 1)   # active = double slit (1)
47  button.on_clicked(updatefig)

49  updatefig(0)
    plt.show()
```

Program A.5.3 Fitting the wave function in a box.

```
    import numpy as np, matplotlib.pyplot as plt
2   from matplotlib.widgets import Slider

4   def updatefig(dummy):        # called when paras are changed
        A, B, n = sliders[0].val, sliders[1].val, sliders[2].val
6       psi = A*np.sin(n*np.pi*x) + B*np.cos(n*np.pi*x)
        plot.set_ydata(psi)
8       ax.set_ylim(-1.2,1.2)
        ax.set_xlim(0, 1)
10      fig.canvas.draw_idle()        # necessary?

12  x = np.linspace(0, 1, 100)        # x grids
    vars = ['A', 'B', 'n']            # A sin(kx) + B cos(kx)
14  vals = [[0, 1, 0.5], [0, 1, 0.5], [0, 10, 5.2]]     # min, max, start

16  fig = plt.figure()
    fig.canvas.set_window_title('Square well potential')
18  ax = fig.add_subplot(111)
    fig.subplots_adjust(left=0.15, bottom=0.35, top=.95)
20  plot = ax.plot(x, 0*x, '-r', lw=2)[0]     # initiate plot
    plt.axhline(0, ls='--')           # y=0 axes
22  plt.text(.2, .8, '$A\, \sin(n\pi x) + B\, \cos(n \pi x)$')
```

```
24  plt.xlabel('$x$ (arb.)'), plt.ylabel('$\psi$ (arb.)')

    sliders = []                        # draw sliders at bottom
26  xw, yw, offset, dx, dy = .15, .20, .05, .5, .03      # widgets pos.
    for i in range(len(vars)):
28      axes = fig.add_axes([xw, yw-i*offset, dx, dy])
        sliders.append(Slider(axes, vars[i], vals[i][0], vals[i][1], vals[i][2]))
30      sliders[i].on_changed(updatefig)

32  updatefig(0)
    plt.show()
```

Program A.5.4 Visualizing quantum states.

```
1   import numpy as np, matplotlib.pyplot as plt
    from matplotlib.widgets import Slider
3   from numba import jit

5   def pot(x):                  # change to your own
        return 0.5*x*x           # make sure to adjust Emin/Emax below
7
    Emin, Emax = 0., 5.          # adjust according to pot(x)
9   R = 5.                       # large enough so wf(+/- R) small
    N = 200
11  m = N//2 + 5          # off-center matching point, must be > N/2
    x, dx = np.linspace(-R, R, N+1, retstep=True)
13  V = np.vectorize(pot)(x)     # vectorize pot just in case

15  @jit
    def compute(E):              # integrate Sch. eqn.
17      wfu = np.zeros(m+1)      # init nonzeros
        wfd = np.zeros(m+1)
19      wfu[1], wfd[1] = 1., 1.
        pfac = 2*(1 + dx*dx*(V-E))
21      for i in range(1, m):
            wfu[i+1] = pfac[i]*wfu[i] - wfu[i-1]      # upward march
23          wfd[i+1] = pfac[N-i]*wfd[i] - wfd[i-1]   # downward
        wfd = wfd*wfu[m]/wfd[N-m]                    # normalize at match pt
25      dfu = (wfu[m] - wfu[m-2])                 # df/dx at m-1
        dfd = (wfd[N-m] - wfd[N-m+2])
27      wfd = wfd[:N-m+1]                         # from the matching pt up
        return wfu, np.flip(wfd), dfu, dfd
29
    def updatefig(dummy):        # called when paras are changed
31      E = slider.val
        wfu, wfd, dfu, dfd = compute(E)
33      err = (dfd-dfu)/max(abs(dfu),abs(dfd))
        scale = max(max(abs(wfu)), max(abs(wfd)))
35      plot1.set_ydata(wfu/scale)
        plot2.set_ydata(wfd/scale)
37      etxt.set_text('E=' + repr(E)[:5])
        dtxt.set_text("$\Psi$ err%=" + repr(err*100)[:4])
39      if abs(err) < 0.05:          # good state 5% level
            num = float(repr(E)[:5])
41          if min(abs(1-np.array(good)/num)) > 0.05:     # a likely distinct state
                good.append(num)
```

```
43      allowed.set_text('Allowed E:' + str(good[1:]))

45  fig = plt.figure()
    fig.canvas.set_window_title('Visualizing quantum states')
47  ax = fig.add_subplot(111)
    fig.subplots_adjust(left=0.15, bottom=0.35, top=.95)
49  ax.plot(x, V/max(abs(V)+1.e-9), 'k-')     # add small num if V=0

51  plot1 = ax.plot(x[:m+1], 0*x[:m+1], 'b-', lw=2)[0]  # initiate plots
    plot2 = ax.plot(x[m:], 0*x[m:], 'r-', lw=2)[0]
53  plt.axvline(x[m], ls=':')                 # match pt
    etxt = ax.text(-R, -.5, '')
55  dtxt = ax.text(x[m+10], .95, '')
    good = [1.e10]
57  allowed = ax.text(-R, -.9, '')
    plt.xlabel('$x$ (a.u.)'), plt.ylabel('$\Psi$ (arb.)')
59  plt.ylim(-1.1,1.1)

61  axes = fig.add_axes([.1, .20, .8, .03])            # energy slider
    slider = Slider(axes, 'E', Emin, Emax, Emin)       # range, init val
63  slider.on_changed(updatefig)
    updatefig(slider.val)
65
    plt.show()
```

Program A.5.5 Quantum computer simulator.

```
import numpy as np
2  from numpy.random import random

4  def Qubit():          # returns spin up [1,0]
       return np.array([1, 0])
6
   def H(qubit):         # Hadamard gate
8      hg = np.array([[1,1], [1,-1]])/np.sqrt(2.0)
       return np.dot(hg, qubit)
10
   def M(qubit):         # Measurement
12     return np.dot([1, 0], qubit)

14 s = 0.0      # spin count
   N = 1000     # samplings
16 for i in range(N):
       q = Qubit()
18     hq = H(q)
       c1 = M(hq)       # amplitude
20     r = random()
       if (r < c1**2): # spin up
22         s = s + 1.0

24 print('Probability of spin up:', s/N)
```

Program A.5.6 Grover search method.

```
1  # See links to online programs below
```

See Chap. 7 Program `p6-grover.ipynb` online at https://github.com/com-py/intro.

A.6   Programs from Chap. 8

Program A.6.1 Nuclear decay.

```
1  from numpy.random import random
   import matplotlib.pyplot as plt
3
   p = 0.05
5  N = 1000
   Nlist = []
7
   while N>10:
9      Nlist.append(N)
       dN = 0
11     for _ in range(N):
           if random() < p:
13             dN = dN - 1
       N = N + dN
15
   plt.plot(range(len(Nlist)), Nlist, '.')
17 plt.xlabel('time (arb.)')
   plt.ylabel('$N$')
19 plt.show()
```

Program A.6.2 Brownian motion.

```
1  import numpy as np
   from numpy.random import random
3  import matplotlib.pyplot as plt
   import matplotlib.animation as am
5
   def move():
7      for i in range(N):
           fac = np.exp(-b*dt)
9          r[i] += v[i]*(1-fac)/b
           v[i] = v[i]*fac
11         phi = random()*2*np.pi              # apply kick
           v[i] += [dv*np.cos(phi), dv*np.sin(phi)]
13
   def updatefig(*args):                       # update figure data
15     move()
       plot.set_data(r[:,0], r[:,1])           # update positions
17     return [plot]                           # return plot object
19 N = 500
   r = np.zeros((N, 2))
21 v = np.zeros((N, 2))
```

```
      b = 0.1
23    dt = 0.1
      dv = 0.002        # impulse
25
      fig = plt.figure()
27    plt.subplot(111, aspect='equal')
      plot = plt.plot(r[:,0], r[:,1], '.')[0]        # create plot object
29    ani = am.FuncAnimation(fig, updatefig, interval=10, blit=True) # animate
      plt.xlim(-.75, .75)
31    plt.ylim(-.75, .75)
      plt.show()
```

Program A.6.3 Energy sharing in a solid.

```
      from numpy.random import random, randint
2     import matplotlib.pyplot as plt, numpy as np
      import matplotlib.animation as am
4
      def exchange(L=20):                        # iterate L times
6         for _ in range(L):
              donor = randint(0, N)              # random pair
8             getor = randint(0, N)
              while solid[donor] == 0:           # find a nonzero donor
10                donor = randint(0, N)
              solid[donor] -= 1                  # exchange energy
12            solid[getor] += 1

14    def updateimg(*args):                      # args[0] = frame
          L = 20 if args[0]<400 else 200         # slower init rate
16        exchange(L)
          plot.set_data(np.reshape(solid, (K,K)))   # update image
18        return [plot]                          # return line object in a list

20    K = 16                     # grid dimension
      N = K*K
22    solid = [10]*N             # 10 units of energy/cell
      fig = plt.figure()
24    img = np.reshape(solid, (K,K))             # shape to KxK image
      plot = plt.imshow(img, interpolation='none', vmin=0, vmax=50)
26    plt.colorbar(plot)
      plt.axis('off')
28    anim = am.FuncAnimation(fig, updateimg, interval=1, blit=True)  # animate
      plt.show()
```

Program A.6.4 Entropy of an atom.

```
1     from numpy.random import random, randint
      import matplotlib.pyplot as plt, numpy as np
3
      def exchange(L=20):                        # iterate L times
5         for _ in range(L):
              donor = randint(0, N)              # random pair
7             getor = randint(0, N)
              while solid[donor] == 0:           # find a nonzero donor
9                 donor = randint(0, N)
              solid[donor] -= 1                  # exchange energy
```

```
11          solid[getor] += 1

13  def entropy():                           # entropy of Einstein solid
        Et, E, n, s = sum(solid), 0, 0, 0.   # Et = tot units of energy
15      pn = []
        while E < Et:                        # until all units are counted
17          cn  = solid.count(n)             # num. of cells with En
            E = E + cn*n                     # cumul. count
19          p = cn/float(N)                  # probability
            if (cn != 0): s -= p*np.log(p)   # entropy/k,
21          pn.append(p)                     # record p
            n = n + 1                        # next energy to sample
23      return s, pn

25  N = 400
    solid = [1]*N                            # initialize solid
27  t, s = [], []
    step = 1
29  for i in range(11):
        si, pn = entropy()
31      s.append(si)
        t.append(step)
33      step = 2*step                        # double steps each time
        exchange(step)
35
    plt.plot(t, s, '-o')
37  plt.semilogx()
    plt.xlabel('Iteration')
39  plt.ylabel('Scaled entropy')
    plt.show()
```

Program A.6.5 The Ising model.

```
    import numpy as np
2   from numpy.random import random, randint
    import matplotlib.pyplot as plt
4   import matplotlib.animation as am
    from matplotlib.widgets import Slider
6
    def flip(E):
8       kT = slider.val                      # kT from slider
        n = randint(0, N)                    # pick one to flip
10      dE = 2*spin[n]*(spin[n-1] + spin[(n+1)%N])   # periodic bc
        accept = False
12      if (dE <= 0.0): accept = True        # flip if dE<0
        else:
14          p = np.exp(-dE/kT)               # else flip with exp(-dE/kT)
            if (random() < p): accept = True
16      if accept:                           # actual flip
            E = E + dE
18          spin[n] = -spin[n]
            xy[n,1] += dy[n]*3.7             # offset y-pos for level arrows
20          dy[n] = -dy[n]
        return E
22
    def updatefig(*args):                    # update figure data
```

```
24      global E, m
        E = flip(E)                      # trial flip
26      plot.set_offsets(xy)             # update pos/dir
        plot.set_UVC(dx, dy)
28      Eavg[m%mc] = E                   # last mc points
        m = m+1
30      etxt.set_text('E='+repr(sum(Eavg)/mc)[:5])   # 5 digits
        return [plot]
32
    N = 32
34  spin = [1]*N                  # initial spins all up
    E = -N
36  mc = 1000                     # running average length
    Eavg = [E]*mc
38  m = 0

40  xy = np.zeros((N,2))          # (x,y) pos of arrows
    xy[:,0] = np.linspace(0,1,N)  # x-pos of arrows
42  dx = np.zeros(N)              # x offset
    dy = 0.02*np.ones(N)          # length of arrow=0.02
44
    fig = plt.figure()
46  plot = plt.quiver(xy[:,0], xy[:,1], dx, dy)
    ani = am.FuncAnimation(fig, updatefig, interval=1)        # animate
48  etxt = plt.text(0, -.1, '', color='black')               # energy
    plt.axis('off')
50  plt.ylim(-.5,.5)

52  fig.canvas.set_window_title('Ising model')
    axes = fig.add_axes([.15, .20, .75, .03])        # kT slider
54  slider = Slider(axes, 'kT', 0.01, 8, 4)          # range, init val
    plt.show()
```

Program A.6.6 Global warming example.

```
1  import matplotlib.pyplot as plt
   from matplotlib.widgets import Slider, RadioButtons
3  import numpy as np

5  sunflux = 1300.
   ocean_depth = 1000.
7  specific_heat = 4182.
   water_density = 1000.
9  mass = 4*ocean_depth*water_density

11 secs = 365*24*3600/2    # seconds in a year, sunny half the time
   time = np.linspace(0, 100, 101) # in years
13
   def updatefig(*args):             # update figure data
15     eta = slider.val
       heat = sunflux*time*secs*(eta/100)
17     if button.value_selected == items[1]:              # nonlinear
           nonl = np.linspace(1, np.sqrt(2+eta/10), 101)**2    # nonlinear effects
19         heat = heat*nonl
       deltaT = heat/(specific_heat*mass)
21     plot.set_ydata(deltaT)
       ax.set_ylim(min(deltaT), max(deltaT))
```

```
23      etat.set_text('$\eta=$'+repr(eta)[:4]+'%')        # 4 digits
        fig.canvas.draw_idle()
25
   fig = plt.figure()
27 fig.canvas.set_window_title('Earth warming')
   fig.subplots_adjust(left=0.15, bottom=0.35, top=.95)
29 ax = fig.add_subplot(111)
   plot = ax.plot(time, [0]*len(time))[0]
31 plt.xlabel('Time (years)')
   plt.ylabel('Temperature change ($^\circ$C)')
33 etat = plt.text(45, .01, '', color='black')          # eta text

35 axes = fig.add_axes([.15, .20, .75, .03])            # trap slider
   slider = Slider(axes, '$\eta$ %', -5, 10, 1)          # range, init val
37 slider.on_changed(updatefig)

39 items = ['linear', 'nonlinear']          # add buttons
   baxes =fig.add_axes([.75, .4, .15, .15])
41 button = RadioButtons(baxes, items, 1)   # active = nonlinear (1)
   button.on_clicked(updatefig)
43
   updatefig(None)
45 plt.show()
```

A.7   Programs from Chap. 9

Program A.7.1 Game of life. Also available in GlowScript.

```
1 # See links to online programs below
```

See Chap. 9 Program ch09.ipynb online at https://github.com/com-py/intro.

Index

© The Editor(s) (if applicable) and The Author(s), under exclusive license to Springer   251
Nature Switzerland AG 2023
J. Wang and A. Wang, *Introduction to Computation in Physical Sciences*,
Synthesis Lectures on Computation and Analytics.
https://doi.org/10.1007/978-3-031-17646-3

Printed in the United States
by Baker & Taylor Publisher Services